PROLIFIC MOMENT

Prolific Moment: Theory and Practice of Mindfulness for Writing foregrounds the present in all activities of composing, offering a new perspective on the rhetorical situation and the writing process. A focus on the present casts light on standard writing components—audience, invention, and revision—while bringing forth often overlooked nuances of the writing experience—intrapersonal rhetoric, the preverbal, and preconception. This pedagogy of mindful writing can alleviate the suffering of writing blocks that comes from mindless, future-oriented rhetorics. Much is lost with a misplaced present moment because students forfeit rewarding writing experiences for stress, frustration, boredom, fear, and shortchanged invention. Writing becomes a very different experience if students think of it more consistently as part of a discrete now. Peary examines mindfulness as a metacognitive practice and turns to foundational Buddhist concepts of no-self, emptiness, impermanence, and detachment for methods for observing the moment in the writing classroom. This volume is a fantastic resource for future and current instructors and scholars of composition, rhetoric, and writing studies.

Alexandria Peary coordinates the first-year writing program and is a professor in the English department at Salem State University, USA, where she teaches courses in creative writing and composition. Peary's blog *Your Ability to Write is Always Present* (www.prolificmoment.com) focuses on mindful writing and has an international following. She has given talks on mindful writing at institutions around the United States and in the United Kingdom.

PROLIFIC MOMENT

Theory and Practice of Mindfulness for Writing

Alexandria Peary

NEW YORK AND LONDON

First published 2018
by Routledge
711 Third Avenue, New York, NY 10017

and by Routledge
2 Park Square, Milton Park, Abingdon, Oxon, OX14 4RN

Routledge is an imprint of the Taylor & Francis Group, an informa business

Library of Congress Cataloging in Publication Data
Names: Peary, Alexandria, 1970– author. Title: Prolific moment : theory and practice of mindfulness for writing / Alexandria Peary. Other titles: Mindfulness for writing
Description: First edition. | New York, NY : Routledge, 20018. |
Includes bibliographical references.
Identifiers: LCCN 2018001889 | ISBN 9781138493575 (hardback) |
ISBN 9781138493599 (pbk.) | ISBN 9781351027663 (ebk)
Subjects: LCSH: Creative writing. | English language–Rhetoric–Study and teaching. |
Mindfulness (Psychology)–Study and teaching. | Attention. | Awareness.
Classification: LCC PE1404 .P3825 2018 | DDC 808/.0142–dc23
LC record available at https://lccn.loc.gov/2018001889

ISBN: 978-1-138-49357-5 (hbk)
ISBN: 978-1-138-49359-9 (pbk)
ISBN: 978-1-351-02766-3 (ebk)

Typeset in Bembo
by Out of House Publishing

For my students

CONTENTS

ACKNOWLEDGMENTS

This book has taken the patience and support of many people—Thomas Newkirk, Keith Hjortshoj, Laura Mullen, Karl Kageff, Bronwyn Williams, Hannah McCarthy, Doreen Demoree, and Rebecca Le Mon at Salem State University, Laura Briskman and Nicole Salazar at Routledge, Georgia Brundige, Kyle Potvin, and Deborah Schillbach. I particularly wish to thank Al DeCiccio for his remarkable generosity of spirit and kind invitations to his campus over the years to speak about mindful writing. To Peter Elbow for his seismic, multi-decade influence on everything I most believe in, I bow deeply in gratitude. I am appreciative of the guidance of the late Teresa Jaros Enos, editor of *Rhetoric Review*, Richard Enos, a reviewer along with Peter Elbow for an earlier version of part of this project, and the anonymous reviewers at Routledge. I am thankful to my students, undergraduate and graduate, who have given their consideration to mindfulness as a part of their writing process and who have demonstrated the courage to enroll in a writing blocks course, sharing their writing journeys of struggle and pleasure. A special bow of gratitude goes out to Marcus Babineau, Claudia Cuneo, Allison Gage, Malissa Pratt, Amanda Richmond, Jonah Sanabia, Jenny Towers, and the student writers whose work appears under a pseudonym for permitting the inclusion of their writing from mindfulness-based first-year composition courses. To my beloved family, Michael, Sophia, and Simone, who tolerated countless hours in which my study door was closed, then opened, closed and opened, the door is now open.

PREFACE

Hymn of Binaries, Mantra for Equanimity: Wooden Sculpture

If you seek something for your writing, allow yourself to be pulled in the opposite direction. Don't resist tides.

So if you seek completion, let yourself be pulled toward the fragmentary, the dissolving, pixels scattering on the horizon, like water receding from stones, like an ellipsis being pulled in, an acknowledgment withdrawn, a closeness evaporating.

If you seek acclaim or acceptance for your writing, let yourself be tugged toward obscurity, let yourself be imprinted with the forks of absent sand pipers.

If you want to write a lot and often, head toward writing nothing, away from the shore and toward that black and white horizon with the numbered cloud.

If you want to write in X genre or on X project, let yourself be pulled toward Y.

If you crave privacy from audience, let yourself be pulled toward full exposure, to immediate performance, and vice versa, if you sorely want to write for an audience, let yourself write for no one.

If you seek to be fully conscious while writing, let yourself be dragged under by the unconscious.

If you wish to forget everything that you have written, remember everything that you have written until the landscape is fifteen or fifty oceans thick.

If you seek to be original, repeat everything twice, three times, for an entire page until the wide-ruled, double lane sea is covered with the same shapes.

If you want to continue your writing session, let yourself stop writing for the day.

If you wish to understand push-pull, let yourself sail along on the hyphen between those two words.

If you want to operate without goals and ambitions, let yourself be loaded with the cargo of those items by the dozens, in car-sized crates, let your ship the size of three football fields be filled with trinkets and non-necessities.

If you prefer to write prose, write poetry. If you prefer to write nonfiction, write fiction.

If you want to spend not so much time at the writing desk, let yourself spend days at a time at the writing desk. If you want to spend many hours at the desk, spend only twenty minutes, then only ten minutes.

If you seek to write free of disturbances, place yourself in a setting in which you will be constantly spoken to.

If you hope to reach destinations of surprise and discovery through your writing, let yourself land on the plateau of nothing new, where the mohawked sun occasionally rests its chin.

And vice versa, reversing the process.

INTRODUCTION

> In your mind, you create an idea of place separate from an actual time.
>
> —*Shunryu Suzuki*

This book explores a theory and pedagogy of mindful writing for the field of composition, an approach that foregrounds the present moment in the activities of composing. No writing is possible without the context of the present moment—whether or not that context is observed is another matter, a matter specifically of mindlessness. This book offers an alternative perspective on the rhetorical situation and the writing process by highlighting present temporality in every writing occasion and at every phase, from prewriting to final edits before audience reception of the last version. The act of writing is located inside an ever-shifting context of a present moment, a situatedness built upon a radical groundlessness that calls for a different treatment of content and a different self-perception. A focus on the present casts new light on regularly discussed aspects of writing—audience, invention, and revision—while bringing forth under- or undiscussed nuances of the writing experience—intrapersonal rhetoric, the preverbal, and preconception, to name a few. Every moment is inventive in the sense that every moment carries traits associated with the generative point of writing: the moment is continuously discursive; continuously changing—more exploratory than organizational; and at a physical distance from audience. A pedagogy and theory of mindful writing can improve how students write and, equally important, how they experience writing. A mindfulness approach can alleviate writing struggle and stress and diminish aversion to writing while bolstering interest, confidence, and fluency. Omitting the present moment of writing will lead to suffering; awareness of the present moment of writing will lead to a more optimal lifelong relationship with writing.

For the most part, we can trace students' success or lack of success with writing to their relation to the present moment during composing. In my twenty-four years as a writing instructor, I have met countless people who struggle with writing, and I doubt I am alone in this experience. Readers of this book have likely encountered dozens of students in their first-year composition courses who dislike, avoid, or battle writing in gradations of discomfort, unease, and insecurity. As Laurence E. Musgrove says, student fear of writing is "a common and tragic cultural attitude toward writing, and one we, as teachers, should address directly in our classes and in our professional organizations" (7). The suffering of writing results from students' inattention to the real-time circumstances of writing, including resistance to the impermanence of the present; stocking of the writing moment with preconceptions about ability, process, and content; and partial intrapersonal rhetoric awareness. This suffering becomes a potent attitudinal constraint in a rhetorical situation and may manifest as a wide range of affective responses, from short-lived irritation, apprehension, and disengagement to helplessness and a complete breakdown in writing performance. I join Robert Boice when he proclaims in his book on writing ease and fluency, "one of my preeminent goals is to bring dark, mindless habits out into the daylight" (*How* 4). It's a long-term problem, since misperceptions begun in the writing classroom often persist after graduation and continue into the professions and a lifetime of writing.

Rarely do I meet people—colleagues, family, friends, or the general public—who are comfortable with writing, despite years of formal education and practice. One doesn't have to search long to find people who admit, often with embarrassment and frustration, that they are unable to write in the way they wish—students who can't turn assignments in by deadline or who profess to dread writing courses, book-less colleagues who worry about tenure, friends who twist themselves into knots dealing with a New Year's resolution to write their novel. Tellingly, the same level of stress doesn't pervade other activities—walking, sitting, or reading—ordinary activities, as ordinary as using words could and should be. Because writing tends to be such a future-driven activity (during which most individuals stray from the present out of fear, dread, ambition, and their training) and moreover a ubiquitous activity, it seems a professional responsibility for writing specialists to rethink current instruction. As a teacher, I suffer when my students don't like to write: when their unaddressed history of past difficulty and fixation on their writing future spills onto the classroom context we share in this moment.

Mindfulness is the observation of ever-changing phenomena as they occur in real time using a detached and non-evaluative outlook. Jon Kabat-Zinn famously described mindfulness as "paying attention in a particular way: on purpose, in the present moment, and nonjudgmentally," and Ruth Baer as "taking an accepting, non-judgmental, non-reactive or non-avoidant stance toward observed experience" (Baer 246; Kabat-Zinn 4). Mindfulness is the modus operandi of Buddhism as much as moving the human body is to dance; it's impossible to act upon Buddhist precepts without this adjustment in our thinking and perceptions.

For this reason, as a "prerequisite" in the Buddhist spiritual system, mindfulness needs to be "cultivated at all stages of an individual's meditative journey" (Shonin et al. vii). As has often occurred in recent decades, mindfulness can be extracted from spiritual belief and used in different contexts as a beneficial approach to thinking and behavior. The general principles of a mindfulness pedagogy for composition are the perception and exploration of the fleeting, impermanent nature of phenomena for the purposes of learning and writing and the acceptance of mental formations (including emotions and the registering of physical sensations) without sorting or attachment—all occurring in real time. Mindful writing uses strategies to engage that impermanence for writing, including a detached, non-evaluative mindset, observing internal rhetoric, and acknowledging verbal emptiness. The emptiness of prewriting serves as the backdrop upon which preconceptions, mind waves, mind weeds, and new content can be observed, monitored, and used for creative-rhetorical purposes. So it is that with mindfulness in general, we triangulate ideas of present temporality, mindfulness, and impermanence to conclude that it takes perception (mindfulness) of what's really happening (the present) to acknowledge change (impermanence) and be released from suffering (nirvana).

For writing courses, by analogy, it takes training in the perception of the present rhetorical situation to acknowledge the constant change in that situation to be released from writing-related suffering such as apprehension, self-doubt, rigid composing rules, and writing blocks. Instead of rhetorical context or rhetorical situation, we might more accurately consider how every act of writing is present-situated and occurs in a *rhetorical moment*. Managing the rhetorical situation involves managing the rhetorical moment, and participating in the event of a writing process means entering it via the present. In first-year composition curricula, the development of mindfulness is synonymous with the development of rhetorical and metacognitive awareness. I make a case for why composition and rhetoric scholars should pay more attention to the present moment in theory and pedagogy, asking, *What is overlooked about writing when we are heedless of the present? What could be the consequences of that oversight? Conversely, what are the benefits of mindfulness for our first-year writing students? How can a first-year writing curriculum help students become more aware of the present for their writing?*

Sustained observation of the present moment for purposes of writing leads to a game-changing mindful invention that pervades every aspect of composing. Every moment of writing becomes an inventive moment. The moment—permissive, exploratory, uncorrected, and multi-faceted—becomes the entire basis for all written work. Mindful invention contributes to the rhetorical tradition its housing of all rhetorical moves, process strategies, and resultant content within the frame of present experience. Mindful invention embodies the three characteristics of invention theory: it is exploratory; it is "an initiation of discourse"; and it includes tacit alongside explicit elements (Lauer 2). Mindful invention is a continuous "phase" that forms the backdrop for the entire writing experience. Every moment is potentially inventive. Since every writing occasion, from prewriting to final edits, occurs during a present moment, every phase in a writing process, no matter how close the text is to delivery to a reader, shows characteristics of mindful invention. Freewriting isn't required for mindful invention, because any method (even aspersed outlining or correcting) which remains aware of the moment will suffice. Writers mindfully invent while prewriting; writers mindfully invent while receiving feedback; writers mindfully invent while editing, and thus writers are able to utilize the benefits of the present writing moment at any time. In fact, mindful invention or the continuous observation of the writing moment has to be this all-encompassing just to ward off non-productive mindlessness. I was tempted to title this chapter "Mindful Invention," but I felt wary of giving the wrong impression. Mindful invention isn't a specially designated, discrete phase in the process of writing. Instead, a perspective of mindfulness changes the timing of the writing process such that it's no longer a simple recursivity of returns and jumps to earlier and later phases in a writing process. Writers who are mindful never return to invention: they remain inventive in a prolific moment.

Addressing the suffering of writing in light of this always-possible invention, the Four Noble Truths of Writing are as follows. The first point (First Truth) is that a student's ability to write is always present. A conscious literate individual can write at any moment, in any place, and through any medium—pen, magic markers, keyboard, voice recognition program, food dye—and to maintain otherwise is a sign of mindlessness. All students don't just possess writing ability: they have *prolific* abilities. Any moment is an inventive moment, and that's precisely why we can all become prolific writers. The second point (Second Truth) is that the present rhetorical situation houses everything a student requires to write—intrapersonal dialog, exigence, rhetorical moves and strategies, the sum in a state of flux. The third point (Third Truth) is that writing difficulties occur because of unawareness of the present rhetorical moment. In order to more fluently write, students need to consistently practice mindful awareness. The fourth point (Fourth Truth) is that mindlessness is the default position to be actively countered through a systematic pedagogy organized around present awareness.

Mindfulness in Higher Education

Mindfulness is increasingly adopted as a teaching practice in higher education. In her 2011 roll call of various academic fields incorporating mindfulness in their curricula, Mirabai Bush lists studio and performance art, religion, chemistry, physics, political science, psychiatry, environmental studies, sociology, law, social work, and behavioral economics (196). A survey of peer-reviewed articles on mindfulness and education in the EBSCO database shows a steady uptick of publications between 2000 and 2014; overall 155 articles have been published on the topic, moving from the single article published in 2000 and peaking at thirty-three in 2012 (Schonert-Reichl and Roeser 4). In a panel on writing program administrator leadership and mindfulness at the 2017 Conference of College

Composition and Communication, Jennifer Consilio described the Mindfulness Across the Curriculum (MAC) initiative she is establishing on her campus, one that is receiving an enthusiastic response from her colleagues. Numerous scholars from across disciplines (certainly too many to mention here) use components of mindfulness to address course-specific learning outcomes. In *Integrating Mindfulness into Anti-Oppression Pedagogy*, Beth Berila demonstrates how mindfulness can help students more compassionately and critically examine internalized oppression and their own values to better engage with challenging social justice issues. In religious studies, Laurie Cassidy deploys mindful breathing to build a feminist counterspace to discuss Christian social ethics and proposes that "contemplative ways of knowing, like mindful breathing, may be a way to retrieve critical thinking" (166). Shelton A. Gunaratne, Mark Pearson, and Sugath Senarath's 2015 *Mindful Journalism and News Ethics in the Digital Era: A Buddhist Approach* explores journalistic ethics through the lens of Buddhism, offering strategies to achieve accuracy and fairness in news coverage along with compassion for a reporter's sources and for self.

The reason for this increasing incorporation of mindfulness across disciplines is clear, since the learning benefits of mindfulness have been enumerated in multiple academic fields. Instructors who incorporate mindfulness in classroom praxis tout its benefits for concentration, critical emotional literacy, and what Arthur Zajonc has called "mobility in thought, the ability to sustain complexity or even contradiction" (Bush 188; Winans; Zajonc 179). Mindfulness emphasizes discovery over rote learning because "[w]hen we practice mindfulness, we are not memorizing what someone else has already discovered, we are setting up conditions in which we can observe the direct experiences in our own minds, bodies, and hearts" (Rechtschaffen 5–6). A sampling of contemporary definitions of mindfulness shows it to be a cognitive practice fostering metacognition (Dan Siegel's "more than just simply being aware. It involves being aware of aspects of the mind itself"); a detachment that readies the learner for critical thinking (Daniel Barbezat and Mirabai Bush's "a healthy relationship with thoughts, so that the student becomes aware of a thought rather than identified with it"); a way to foster the well-being of students by returning the affective dimensions to learning and countering "value-impoverished technicist and economistic" competency-based educational systems (Terry Hyland); and for new ideas and perceptions (Mary Rose O'Reilley's "the practice of mindfulness allows us a chance, at least, of seeing the pure data") (Barbezat and Bush 123; Hyland 23; O'Reilley 10; Siegel 5–6).

In her multi-decade, collaborative research on mindfulness, Ellen J. Langer exposes the extent to which traditional educational practices actually undermine genuine learning through their most basic assumptions about what it means to learn. Langer problematizes fixed values such as paying attention, focus, mastery, and practice by demonstrating how they are attributable to automatic behavior and learned passivity. In addition to the typically esteemed learning behaviors, Langer criticizes the kinds of material usually taught—fundamentals, general principles, and categories—for potentially undermining creative or critical thinking, causing habitual rather than innovative thinking. Because people mistakenly believe that cognitive and creative possibility is limited, they put stock in static categories which further underestimate that possibility "rather than accept[ing] the world as dynamic and continuous" (*Mindfulness* 27). In turn, those categories enable people to "make rules by which to dole out these resources. If resources weren't so limited, or if these limits were greatly exaggerated, the categories wouldn't need to be so rigid" (*Mindfulness* 27). Students are disempowered by the mindlessness of traditional education when they fail to perceive the divergences and variations possible in information; misperceiving their resources as limited, students run the risk of a learned helplessness ("A Mindful Education" 44–45). According to Langer, the consequences of this mindlessness run from minor to serious assumptions both in and out of the classroom, ranging from minor faux pas like talking to a store mannequin to jumping by accident into a drained swimming pool (44).

In Langer's work on mindfulness, context is key. Langer's disciplinary lens is social psychology, which examines an individual's context to understand individual behavior: an emphasis on context that she transposes onto learning behavior. Contexts are a matter of individual subjectivity—established in the scene of learning by the person's perceptions—rather than something external to the person (*Mindfulness* 36). Contexts become overly familiar through repetition and an outlook geared toward mastery and expertise; individuals essentially are desensitized to the particularities of a current context. Premature cognitive commitments occur when a person reuses a mindset, trucking it from an originary context into another situation because the transplanted context is erroneously perceived as similar and familiar (*Mindfulness* 22). Overlearning, for instance, results when a student has practiced to such an extent that the student operates on generalizations about unique contexts. In sum, what's installed through standard teaching methods is automatic behavior rather than a critical—or mindful—consciousness. The second mishandling of context and subsequent source of mindlessness happens when an individual doesn't just transpose a former context onto a current context but actually erases context: "When mindless, however, people treat information as though it were *context-free*—true regardless of circumstances" (*Mindfulness* 3). As a remedy, Langer maintains that mindfulness or the perception of context promotes more active, conditional learning, helping students avoid the trap of easy and early certainty (*Mindfulness*). According to Langer, a "mindful approach to any activity has three characteristics: the continuous creation of new categories; openness to new information; and an implicit awareness of more than one perspective" (*The Power* 4). A mindful approach to learning supports novelty or the search for variations instead of relying on established categories and instead of "acting from a single perspective" (*Mindfulness* 16). At the same time, Langer adopts a reasonable view toward mindlessness. It's not altogether avoidable; we need to work with provided systems and structures, beginning with the alphabet, so a certain amount of autopilot or rote thinking is inescapable, and a mindless response is warranted if it is most appropriate for a situation that does not significantly change (Langer and Piper 285). This view is not only pragmatic but, in my mind, in synch with Buddhist practice, since it allows two entities, in this case mindfulness and mindlessness, to be in interplay rather than be kept dualistically separate, and as I will shortly discuss, mindful composition in fact makes room for mindless experience.

In recent years in Composition Studies, mindfulness has been discussed in relation to a host of composition-related issues, emerging from a larger, multidisciplinary movement toward contemplative educational practices in the United States. This movement has evolved several dedicated journals, including the *Journal of Contemplative Inquiry*, the *Journal of the Assembly of Expanded Perspectives on Learning*, and the *Journal of Transformative Education*, an annual national conference, and at least one list-serv. As defined by The Center for Contemplative Mind in Society, contemplative practices are "practical, radical, and transformative, developing capacities for deep concentration and quieting the mind in the midst of the action and distraction that fills everyday life." Contemplative practices include deep listening, meditation, aikido, yoga, journaling, volunteering, vigils and marches, lectio divina, beholding, pilgrimages to sites of social justice, retreats, and improvisation ("Tree of Contemplative Practices"). In Composition Studies, mindful composition is showing signs similar to those in the process movement when it made its renaissance to a pedagogy: mindfulness is incorporated in a wide range of contexts and has a burgeoning body of scholarship (Tobin 6–8). Already documented in the growing body of mindful writing scholarship are individual classroom approaches as well as rhetorical analysis, for instance, of ancient Buddhist sutra.

For the composition classroom, states of mindfulness (and mindful activities such as yoga) are equated with invention and critical thinking. Scholars propose that mindful awareness helps students transition from the preverbal to invention; avoid preconceived thinking and improvise; reach insights that bolster students' rhetorical engagement, meeting key instructional outcomes of creativity, listening, and expression; and develop self-acceptance of their writing (Crawford and Wilhoff; Forrester; Kirsch "From Introspection"). In the 1960s, meditation was linked with prewriting, along with journal writing and analogic thinking, as a method to help students with the discovery phase of writing (Rohman). Article titles with "meditation" showed up in the 1980s, such as James Moffett's "Writing, Inner Speech, and Meditation" and "Reading and Writing as Meditation"—in which he explored "discursive meditation," or the ways in which meditation can direct attention to and silence inner speech for the classroom. In "The 'Not Trying' of Writing," Rachel Forrester, guided by Barrett J. Mandel, describes how writing is an ongoing but rewarding struggle if approached with a Buddhist stance of acceptance and non-effort (53). Irene Papoulis posits an inherently spiritual component to composition instruction because of the transition from preverbal to invention that happens whenever someone embarks on writing. Ryan Crawford and Andreas Wilhoff describe how meditation in a composition classroom can help students disinhibit, avoiding preconceived thinking, gaining insight, improvising material, and increasing intrinsic motivation by creating an outlook of stillness during a writing task. Mindful physical presence has been correlated with invention in a somaesthetics of composing (Cohen; DeLuca; Fleckenstein *Embodied*; Perl; Wenger, "Writing Yogis" and *Yoga Minds*; Wilson). Robert Tremmel's *Zen and the Practice of Teaching English* offers four pillars of teaching based on Basho and extends mindful attention to the everyday activities of the classroom. A recent publication, Barry M. Kroll's *The Open Hand: Arguing as an Art of Peace*, is concerned with how martial arts and mindfulness facilitate noncombative and cooperative instruction in argument. The several benefits of mindful awareness for this handling of conflict include how it "allows one to notice that an argument is arising, that certain emotions are being generated, and that various intentions and strategies can be mobilized, depending on the situation" (Kroll 15–16). Mindfulness is among the three strategies Elizabeth Wardle recommends to build transfer through writing-about-writing, albeit with the qualification that mindfulness can be encouraged but not forced upon students—odd, since she doesn't apply the same disclaimer to reflection or abstracting strategies (771). (Indeed, the Buddha developed an extensive and long-lasting pedagogy to teach metacognition, as I discuss in Chapter 1.) Recently, Ellen C. Carillo proposed mindful reading in first-year composition courses, whereby students examine their immediate reading context to select the most applicable reading strategy for the task at hand, switching reading methods as warranted. Students "reflect on the present moment [to determine] how far a reading approach takes them, what aspects of the text it allows them to address, and what meanings it enables and prohibits" (124). Mindfulness here functions as metacognition, an overarching critical perspective in which students detach, step back, and observe present occurrences in their reading and consciously select from a "repertoire of reading approaches they have been cultivating in first-year composition" (124).

In *Yoga Minds, Writing Bodies: Contemplative Writing Pedagogy* and an earlier article in the *Journal of the Assembly of Expanded Perspectives on Learning,* Christy I. Wenger grafts the contemplative practices of yoga and yogic breathing into first-year composition courses to foster "mindfulness of the body," a somaesthetic through which students gain metacognitive skills about rhetoric and process (*Yoga* 11). Drawing from the feminist theorist Donna Harraway and yoga pioneer BKS Iyengar, Wenger pursues greater integration of mind and body, self and other, in student writing experience, and helps students unlearn disconnection from their bodies. Wenger's term for that integration is "embodied imagination," or "the faculty by which body, heart, and mind work together to bring meaning and understanding to writing" (*Yoga* 21). While most resources on mindfulness in composition instruction handle mindfulness as a learning outcome rather than a heuristic and thus position mindfulness as the result of activity—*calm, reflective, open-minded*—and not *as activity*, Wenger's work is an exception. She treats mindfulness "as both a heuristic for contemplative pedagogy and a body-minded habit achieved through consistent involvement in contemplative practice" (*Yoga* 105). This pedagogical approach to mindfulness is aligned with traditional Buddhist pedagogy, in which mindfulness is more a disciplined practice than a quality to obtain. Indeed, the historical sense of mindfulness casts at least as much emphasis on cause as on effect; mindfulness is a translation from Pali of *sati*, a word which made mindfulness both the purpose and the means of meditation (Gunaratana 137; 145). As Paul Grossman and Nicholas T. Van Dam explicate this difference, *sati* is "perhaps best translated 'to be mindful,' in stark contrast to the use of the word 'mindfulness,' which is, of course, a noun and easily implies a fixed trait" (220). The active nature of Wenger's approach is evident in her comparison to mindfulness as an intervention—the connotation of which suggests purposeful, swift, and even daring action. Mindfulness is an "embodied intervention that creates a rich source of practice and theory which can be used to transform the work completed in our college writing classrooms and the ways that work is transferred to other writing environments" (*Yoga* 5). Yogic breathing or *pranayama* specifically shows students how to use their physicality to gain insights about their writing process and develop resilience—Wenger calls this "emotional flexibility"—in the face of unpleasant feelings arising from writing activity. Wenger says that "as students breathe their way into writing, they place new value on observing the writing process as it unfolds" and "students' increased mindfulness and flexibility results in developed focus and advanced coping mechanisms to deal with the negative emotions of the writing process" ("Writing Yogis" 26–27).

As an embodied epistemic, Wenger's mindfulness works to unknot the stubborn binary of interior/exterior or self/social that has preoccupied the field of composition for decades, easily since the work of James Berlin. Awareness of our particular embodied experiences (her term for this is "presence") doesn't lead to solipsism or inflate our own subjectivities; instead, it can highlight our connections to others through their own materiality (her term is "resonance"). As Wenger says, "a lived, moment-to-moment understanding of materiality ... allows us to approach a writer's agency as singular, situated in a particular body, and located via her interaction with other material bodies—even if it is also social" (29). Indeed, observing the changing moment means observing material conditions. For instance, if I am mindfully observing my breath right now at my desk, I will notice the seat of my chair, an object designed, produced, marketed, and shipped by innumerable faceless others, gifted to me by a long-absent friend, not to mention the dozen or so textual productions of other authors stacked at two corners of my desk or taped to the wall. When we do this sort of reflection, we invariably notice physical sensations which are not attributable to us but, rather, are the results of other people's material conditions and actions—one of the provisions to composing afforded by yogic mindfulness.

Interest in mindful writing has also flourished in the extracurriculum—frequently a harbinger of major developments in institutional academia, as Anne Ruggles Gere and others have indicated. Mindfulness and writing have already attracted significant interest outside academia, evinced in Natalie Goldberg's best-selling *Writing Down the Bones* (1986), which adopts a Zen approach to writing, but also Dinty W. Moore's (2012) *The Mindful Writer: Noble Truths of the Writing Life*, Brenda Miller and Holly J. Hughes' (2012) *The Pen and the Bell: Mindful Writing in a Busy World*, and Gail Sher's (1999) *One Continuous Mistake: Four Noble Truths for Writers*. Typically, these trade books are top-heavy on exercises meant to guide the self-learner and use anecdotal evidence or lore rather than a theoretical-historical foundation. They offer craft advice or the how-to wisdom exchanged between a seasoned writer and a less experienced one, based on what has been observed to work for one person. Regardless, the existence (and success) of self-help books on mindful writing speaks to a sizeable interest in the development of writing abilities through mindfulness. While I join others in their caution about popularized, overly simplified applications of mindfulness, suffice it that this group of self-help books joins a spate of other trade books in which mindfulness has been applied to a large arena of human activities—dieting, management, psychology, dating.

Mindlessness in Writing Instruction

Currently, a lack of present-moment awareness riddles composition praxis despite our collective intention to do well by students. To an extent, it's a state of affairs to be expected given the human propensity, no matter what the profession or activity, for mindlessness. Most of us can't sustain a now-oriented focus on or off the writing desk without pretty quickly departing into our intrapersonal rhetoric, which takes us off-roading from the moment into future- or past-based hypothetical considerations. As Thich Nhat Hanh describes this conundrum, "We began to run a long time ago. We even continue to run in our sleep. We think that happiness and well-being are not possible in the here and now. That belief is inherent in us. We have received the seed of that belief from our parents and our grandparents ... That is why when we were children, we already had the habit of running" (*The Path* 21–23). The typical human attention span for the present was estimated in the 1880s by Wilhelm Wundt to be between five and twelve seconds when he tried to measure "the duration of the present—that interval of time that can be experienced as an uninterrupted whole" (Kern 82). In *Principles of Psychology*, William James claimed that the average experience of the present moment was limited to a "vaguely vanishing backward and forward fringe; but at its nucleus is probably the dozen seconds or less that have just elapsed" (613). Humans demonstrate a proclivity toward avoiding the present: we can't sustain a now focus without departing into "monkey mind" or inner discursivity concerned with evaluation, the past, or the future.

It's almost certain that the very act of writing—no matter what the genre or audience—compounds our mindlessness problem by asking us to produce texts in the now for an expected future situation in which we're not likely to be an on-site participant (and are therefore unable to defend or explain ourselves). As Peter

Elbow points out in *Vernacular Eloquence*, writing comes with physical modalities of both time (unlike the ephemera of unrecorded speech, it persists) and space (it occupies a physical dimension)—qualities I believe warrant close monitoring for latent mindlessness (14). Student writers constantly cope with the rhetorical and temporal disconnect of preparing a missive in one passage of time that will be seen by readers in an altogether different setting and at a different date. It would seem unavoidable that students will anticipate and want to prepare for this impending event. As Keith Hjortshoj says in *Understanding Writing Blocks*, "In writing, the act of utterance and the act of communication (or performance) are separated both in time and in space. And this separation represents both the great disadvantage and the great advantage of writing—the cause of its difficulty and the source of its power" (20). Indeed, for most writers at least some of the time, what makes writing seem worthwhile is its promise of an impact on or connection with people in the future.

Not only do the stressors of writing seem to occur because of the future: the benefits of writing also seem to occur in the future. They're promises of payment for our efforts, so we tolerate the uncertainty, isolation, and other challenges of working on a piece of writing because of its apparent long-term consequences. Students tolerate a complex relationship with teachers on behalf of the long term in the form of grades, graduation, and even growth. As a result, writing seems to contract out the future, because it says it will put us in touch with others, allowing us to express, inform, persuade, entertain, and impress. The act of writing may imply a future for our efforts by telling us we have a responsibility to others because of our words. We need to deal with the chimerical nature of writing when it's stressful as well as when it's alluring. In either form, this mirage leads writers to shadow box with audiences and treat even their earliest starts as public performances in a mindlessness that's a disengagement from the present rhetorical situation and all its affordances. Unfortunately, the field of composition has largely responded to the problems of this disjuncture through denial, a mindlessness that only increases student stress.

On the face of it, written assignments seem less susceptible to rote learning and memorizing lecture material; nevertheless, it is common for students to tolerate blind spots in their awareness as they write. Writing only appears to be less susceptible to mindlessness, since a written piece is never a duplicate of a prior document (an earlier draft or library research) and so inherently includes variation and change. All the same, writing happens on autopilot whenever students are unaware of the moment in which they write. Many individuals inside and outside of classroom settings operate from a single perspective because they become involved in illusions in which the text is already finished and is already readable by an as-of-yet-nonexistent reader. They trick themselves into adopting a single static form. Although the actual draft at hand is under construction and its appearance still changing, the fantasy text of the future is an unchanging object that annexes their thoughts.

Moreover, this imagined text is constructed from a series of premature cognitive commitments made about the future, assumptions likely generated by categories and inherited structures not based on data from the actual moment. These assumptions about the fate of a piece of writing often involve the transposition of a prior writing experience, especially its reception or evaluation, onto the current context. Attempting to reach a fixed, finished document, writers treat a draft in light of an imagined final version and consequently forfeit invention and the continuous creation of new possibilities. Similarly to what Ellen Langer says of learning overall, students studying writing should be taught to look at their immediate context, the one actually at hand. It's a double bind, because in addition to working with the dubious material of assumptions, in this typical scenario of mindlessness, a student also forfeits richly contingent, ever-shifting, dynamic material contained in the impermanence of the present moment.

In fact, writing as it is largely taught, from the earliest to the most advanced years in schools, is one of the main perpetuators of trained mindlessness in people's lives. Widely accepted practices of writing curricula override present-moment awareness: demanding constant consideration of factors beyond the present writing situation, most readily the mandate that students "consider audience" sometimes as prematurely as in the prewriting phase; the future tilt of assessable writing products and learning outcomes; the overwhelming emphasis on interpersonal communication at the expense of the intrapersonal; the general omission of embodied and affective experiences pertaining to writing—all are a prescription for non-beneficial mindlessness. Robert Yagelski, concerned about Cartesian dualities in mainstream writing education, regrets how

> we teach separateness rather than interconnectedness; we see a world defined by duality rather than unity. As a result we promote an idea of community as a collection of discrete, autonomous individuals rather than a complex network of beings who are inherently interconnected and inextricably part of the ecosystems on which all life depends. These lessons are encoded in the institutional structures of schooling, the mainstream curriculum at both the secondary and postsecondary levels, and conventional pedagogies, including writing pedagogies.
>
> *(17)*

Yagelski wants to replace the mindlessness of mainstream education with an ontological focus on writing as a way of being, asking: "[w]hat if we were to refocus writing instruction on the act of writing in the moment, rather than on the production of specific kinds of texts that are valued in academic settings, as mainstream writing instruction does?" (170). The problem of the mindlessness of writing instruction is further compounded by the magnitude of students' exposure to those mindless tendencies. In the fall of 2016, 50.7 million students attended public schools in the United States, and 15.1 million of those students were in high school, presumably taking increasingly sophisticated writing courses (National Center for Education Statistics). In 2010, a conservative estimate of the number of sections of first-year composition courses was 67,932 at twenty students per section, with another estimate putting the number of first-year composition students at around 2 million a year (Hesse; Yagelski 2). Fortunately, the ubiquity of writing courses in individuals' elementary to postsecondary lives presents a tremendous opportunity to change this situation, and first-year composition courses, as instructional sites for process and rhetorical training, are unmatched in this regard. With even minor modifications, several practices familiar to a first-year composition classroom can serve as vehicles for mindfulness—informal writing, disposable writing, freewriting, and feedback, to name a handful—and the metacognition of process and rhetorical awareness is already a step in the direction of mindful awareness.

For instance, an interest common to all instructors of first-year composition in the United States, no matter what their institutional affiliation—two-year, four-year, Ivy, or public—is that students avoid habitual, mindless, and repetitive approaches to writing. As writing instructors, we want our students to attune to the particulars of writing context so they don't manhandle every writing task as a five-paragraph essay or drag around an unexamined attitude about writing from one assignment to the next. Metacognitive skills, or the ability "to notice changes to ways of thinking and to adapt execution of the writing process in order to respect these new ways of thinking," are integral to writing instruction (Wenger *Yoga Minds* 121). Metacognition is everywhere evident in position statements in our field—featured prominently as one of the seven habits of mind in the 2011 "Framework for Success in Postsecondary Writing" developed by the National Council for Teacher Education (NCTE), the Council of Writing Program Administrators, and the National Writing Project—and its variants of meta-awareness and reflection make regular appearances in discussions of transfer, the purpose of a writing education, and Writing About Writing. As Kristine Johnson points out, metacognition in the guise of reflection becomes a prevalent disciplinary concern, since it is "often defined broadly and powerfully, actually encompassing—and perhaps obscuring—habits of mind that the *Framework* outlines as distinct intellectual practices," a sort of disciplinary kudzu (525). First-year composition pedagogy is replete with different goals of awareness: our students regularly practice rhetorical awareness, genre awareness, audience awareness, disciplinary awareness, and process awareness.

This mandate for metacognition, however, becomes problematic when it works counter to its own goals, which it does whenever that metacognition is cast, as it often is cast, as a matter of retrospection, a past-tense rather than present-tense experience. Metacognition frequently occurs after the fact, such that composition students are asked to reflect afterwards on moves made during an already experienced writing process rather than being shown how to observe their ongoing, present rhetorical situation. Mainly, this becomes an issue in two regards.

First, it impairs learning, as most behaviors of mindlessness are apt to do, by reducing openness of perception and the ability to recognize divergences and variances possible in information, such that by misperceiving their resources as limited, students run the risk of a learned helplessness. Contexts become overly familiar, students become desensitized to the particularities of a situation, and premature cognitive commitments ensue. Those premature cognitive commitments occur when an individual reuses a mindset, transplanting it from an originary context to another situation because the transplanted context is erroneously perceived as similar and familiar (Langer *Mindfulness* 22). Since the *Framework* purports to support learning in general vis-à-vis writing, such that the eight habits of mind (curiosity, openness, engagement, creativity, persistence, responsibility, flexibility, metacognition) occur through provisions of writing (rhetorical knowledge, critical thinking, writing processes, knowledge of conventions, ability to compose with different technologies), this omission of present-based experience is concerning.

In most approaches to metacognition and composing, you get the sense that we're standing around the writing moment, hoping that some writing good will come out of taking that position, without exactly knowing what to look for in the present rhetorical situation—or how to teach it. These accounts generally lean on the word "as" to suggest a writing event unfolding in real time, such as the view that students "must understand how to take advantage of the opportunities with which they are presented and address the constraints they encounter as they write" from the 2015 Conference on College Composition and Communication (CCCC) position paper, "Principles for the Postsecondary Teaching of Writing," a point we will explore further in Chapter 5. Then there's the basic nature of metacognition: it's the observation of thoughts and feelings as they unfold in real time. As Michael Lamport Commons and Dristi Adhikari have said, mindfulness "is the practice of actively noticing where one's attention is placed to be in the present" (193). If we're developing awareness of mental formations from the past, that's actually reflection, and if those formations are about the future, that's actually projection. It's possible that the omission of the present in composition praxis as reflected in national writing policies like the *Framework* belies metacognition and makes writing more problematic for students.

While traces of present temporal awareness can be found in composition theory and praxis, as I will discuss in the next chapter, by and large, what Richard Enos said of classical rhetoric applies to the field of composition at large: omitting the "momentary," scholars study "situational constraints" but fail to "capture well the notions of immediate time in the dynamics of composition" (78–79). An example of a perplexing view of present temporality occurs in Frank Smith's handling of metacognitive awareness. Smith makes awareness an atemporal matter, saying that "thought develops through time, but awareness has to stop time" and this freezing of time limits activities of awareness to retrospection (45). An oxymoron, this awareness means a complete detour from what's happening right now. Smith adds that it's this backwards-looking awareness that frequently leads to problems for writers, "probably a sign that we are having difficulty in formulating something that we want to write or in comprehending what we are trying to read" (45–46). In composition theory, most treatment of present temporality is not as overtly mindless as this. Instead, our understanding of the time involved in writing simply skids to a stop startlingly short of arguably the most fundamental part of writing—the present moment—and bypasses it even in well-established constructs that would suggest temporal consciousness such as "writing is a process." Even contemplative pedagogy needs more of a mindfulness perspective when it concerns the educational bedrock of writing: Daniel Barbezat and Mirabai Bush's 2014 *Contemplative Practices in Higher Education*, so cogent in other regards, provides short shrift on how to teach mindful writing: just a few pages covering journal writing and advice about frequency of writing.

A Mindfulness Practice of Composition

What, Then, Is the Nature of This Composing Present?

Fleeting and ongoing, the present moment provides content based on its transience. The present also serves as a forum for the preverbal emptiness that invites the intrapersonal; it is the home base of the internal rhetoric that, in turn, supports any interpersonal communication. Informed by the social, the present rhetorical moment is intertextual with secondary presences of others through print, digital texts, phones, recalled conversation, overheard conversation, email, and any other text visible in a student's working space. The present of writing is a full-body experience that includes awareness of posture, temperature, pain, pleasure, and sensory experiences from clothing, jewelry, and shoes. Finally, the present of composing is moderated by the metronome of the breath, a quintessentially portable, open-access tool: *breathing in, I am aware that I am writing; breathing out, I am aware that I am writing.*

A pedagogy of mindful composition enhances student understanding of writing struggle by demonstrating how intentional, sustained awareness of the present can lead to fluency and ease for students at any level, including first-year composition students. Negative writing experience has been a focus in the work of Peter Elbow, Mike Rose, Robert Boice, Keith Hjortshoj, John Lofty, Alice Flaherty, John Daly, and others. Noble work has been done by Robert Boice, in his study of writing unease and on strategies for establishing a calm outlook during writing, and by Keith Hjortshoj, who reminds us that writing occurs in time as a "psycho-physical" action, a "kind of embodied movement, more or less coordinated by ideas about what we are and should be doing" that occurs in time (10; 11). In *Understanding Writing Blocks*, Hjortshoj explains that writing is not merely a mental activity but a combination of a mental and a physical activity, and thus, writing problems can be resolved by a more accurate perception of a student's "movements within the writing process" (3; 11). Mike Rose contributed a cognitive approach to an array of approaches to student writing blocks that included the "behaviorist (to explore histories of unpleasant writing experience), psychoanalytic (to explore deep-seated fears and defenses), and sociological/political (to explore the environmental conditions that limit at writer)" ("Writer's Block" 2–4). However, in theories of writing blocks, the situation of struggle has been ascribed to only certain student populations. Rose associates blocks only with students of advanced skills, examining when a "capable writer cannot write" ("Writer's Block" 1), rather than developmental students. Hjortshoj suggests that "[y]oung inexperienced writers rarely encounter serious blocks" because blocks happen with sustained projects like dissertations, not with the shorter assignments of a composition course (3). Yet one of the most striking examples in *Understanding Writing Blocks* (other than an aside about a poet who donned an ugly bathing cap to distance herself from imagined audiences) is that of a student he calls Paul, who resolved an onerous writing block during a first-year writing course (14–17). On this point, I am aligned with Lee Ann Carroll, who thinks that "developing meta-cognitive awareness … is useful for students who already know 'how to write' as it is for less well-prepared writers" (38). This awareness can reduce the preconceptions skilled writers maintain about the writing moment, because lacking "such awareness, 'good' writers may find it especially difficult to change writing strategies that have worked for them in the past" (121).

I propose that we replace "writing block" with "mindless writing," a term which allows for the Janus-faced nature of mindlessness and mindful writing and the way mindlessness can act as both non-productive blocking and the creative mindlessness partially akin to a state of flow. Mindless writing suggests the condition of impermanence and how no situation is static, how writing is only a moment away from not-writing, and vice versa. This brings us to one of the most important paradoxes of mindful composing: it takes writing *mindfulness* to reach writing *mindlessness*. Mindlessness is a fundamental part of mindful writing: even this binary of awareness/lack of awareness is subsumed inside a radical groundlessness. Mindlessness and mindfulness coexist; a writer's goal is, to no small extent, to slip into a state of absorption where one idea after another draws us peacefully along on an intrapersonal verbal flow, down a passage of energy and thought to the pleasure of productivity and accomplishment. We're handed a stretch of freedom from our doubts about writing, and our efforts are accompanied by the happy rattling of the keyboard—a state of engagement so consuming that writers are unaware of the passing of time, contentedly lost in the work. Productive mindlessness is similar to how Mihaly Csikszentmihalyi describes flow in terms of its causes and the qualities flow lends to human experience. Both Csikszentmihalyi's flow and productive mindlessness for writing happen when a person is able to maintain a moment-by-moment awareness of his or her consciousness, and both lead to what Csikszentmihalyi calls an "optimal experience" of happiness and enjoyment (3).

What differentiates productive mindlessness, however, from flow is that with this mindlessness any meta-awareness of present developments fades into an often delightful oblivion that is only possible through the support of prior moments of present awareness. In other words, productive mindlessness is a type of post-mindfulness. Dharma legend has it that the poet Allen Ginsberg was slapped on the shoulder by monks for keeping a notebook and pen beside his meditation cushion during group meditation sessions. One reason why the mindful composing theory I am proposing is not synonymous with Buddhist meditation is that this theory is first and foremost a writing strategy—with the goal of writing, and of improving the conditions of writing for students, in mind—whereas a meditation practitioner eschews such an agenda. The problem, however, is that the capacity of the intrapersonal to float us along is finite, and we will be summoned back to our awareness of the moment, of where we are, of what we are doing, of the fact that we want or need to write. Possibly, we will be beached upon our longing for a return of that state of seamless, oblivious productivity. It is at that moment that individuals most require mindfulness, because by returning our attention to the present writing moment, we will almost immediately be able to submerge ourselves in that intrapersonal rhetoric—to find mindlessness writing.

Let me be clear about this point: most of us are functional *mindless* writers. The majority of writers are mindless, and the brunt of writing is mindless writing: immeasurable paragraphs, chapters, theses, poems, freewrites, dissertations, treatises, genre analyses, novels, comparison/contrast papers, research papers, literacy narratives, five-paragraph essays, reviews, cover letters, and so on have been composed out of near to fully mindless states. Innumerable documents have been composed with not a single millisecond of present awareness, a situation which has been happening for generations. What I am arguing most adamantly in this book is that this mindlessness has occurred at a high cost to writers' self-confidence, self-efficacy, and enjoyment as well as a loss of possibility.

The first challenge of mindfulness for writing is to "remember to remember"—to remember to notice the present moment, a skill which cannot be assumed and needs to be taught. Reminding was the agenda of the word *sati*, an important term that predated "mindfulness" in historic Buddhist texts. Sati is translated as "memory" or "to remember" but differs from our current notion of those two terms in that it refers to the continuously "calling-to-mind, being-aware-of certain specified facts," chiefly the fleeting nature of all phenomena (T. W. Rhys David, qtd. in Gethin "Definitions" 265). According to Steven Stanley, mindful awareness is an "embodied and ethically sensitive practice of present moment recollection" (65). The contemporary Vietnamese Buddhist monk Thich Nhat Hanh has also said that "[m]indfulness is remembering to come back to the present moment," a sense captured in the Sanskrit word *smriti* for mindfulness (*The Heart* 64). In monasteries and meditation centers, it is not uncommon for a bell to be periodically rung to call practitioners back to the moment, and householders are advised to pick a frequent, ordinary activity, like using their turn signal or touching a door knob, as their reminding bell. I concur with Erec Smith when he says in "Buddhism's Pedagogical Contribution to Mindfulness" that students need to be shown coherent ways to practice mindfulness and that intellectual connections between writing and mindfulness will not suffice. Of the parallels between *kairos* and mindfulness, for example, Smith agrees that they "may give clues to rhetoric's usefulness in acquiring mindfulness, but such knowledge does not necessarily make that acquisition easier." His intent is to "help students gain mindfulness by avoiding mindlessness [and] acquire the ability to embrace the present as fully as possible" (38). As Anālayo explains, "mindfulness is something that needs to be intentionally brought into being" as a practice of "thorough attentiveness" (31; 33). One of the main charges of a mindful writing pedagogy for first-year composition is to help students remember to notice the moment and transition from remembering to implementing the contents of that moment into their writing.

In the classroom, a mindfulness approach provides overt and ongoing instruction in elements of writing including intrapersonal rhetoric, affective dimensions, flux, detachment, and emptiness. Nothing about mindfulness is presupposed given the human proclivity toward mindlessness, at or away from the writing desk. Instructors shouldn't assume, for instance, that first-year students know to consistently direct attention to their breathing, so techniques should be demonstrated during class time and followed by assignments that allow students to practice and discuss the experience. As an ongoing component of instruction, mindfulness can't be a one-time fix or a single lesson in present-based awareness. It's imperative that both I and my students observe the present throughout the semester while studying the writing process, rhetoric, and genre. Otherwise, mindless habits infiltrate each new rhetorical situation and operate as potent constraints outside the radar of the unwitting writer. The notion of a rhetorical situation, conventionally framed as exigence, audience, and those externally derived constraints such as word count, a deadline, or disciplinary conventions, lacks the direct metacognition of mindfulness in which student writers track and analyze their thinking. In a mindfulness praxis, the quantity/quality equilibrium tilts to quantity, as quantity means tracking the intrapersonal flux and moderating evaluative tendencies. Mindful listening is taught in conjunction with giving and receiving feedback, so that implementation of others' suggestions demonstrates the intertextual nature of writing, interbeing, and the permeable border between ego and other. Inventive revision techniques model how verbal emptiness occurs as students work with drafts, returning to formlessness out of form, or return to the more fleeting and fragmentary qualities associated with prewriting.

Finally, those involved in advocating mindfulness as an educational practice have weighed the risks of appropriation when mindfulness is utilized in a classroom. Tied to learning outcomes, not to mention its goal of helping students break free of writing struggles to reach states of flow, mindfulness used in composition instruction might seem to contradict the very spirit of non-attainment of a mindfulness practice. It may appear to hazard "spiritual materialism" or the trap Chögyam Trungpa described of using spiritual experiences to bolster the ego (*Cutting* 3). Mirabai Bush asks of the emerging mindfulness-across-the-curriculum movement, "The question of instrumentalism. How do we teach students that cultivating mindfulness is likely to have certain beneficial outcomes while helping them to practice in the spirit of open exploration?" (196). Speaking of writing instruction, JoAnn Campbell mused, "It's perhaps a particularly capitalist perspective to think of meditation [in the classroom] as a means to an end" (251). The solution that has been reached by many interested in mindfulness in settings outside a spiritual practice is that mindfulness can and should be used as a pragmatic application in which, as Rupert Gethin has said of the rise of Mindfulness-Based Stress Reduction and Mindfulness-Based Cognitive Therapy, "what is emphasized is the therapeutic usefulness of mindfulness rather than its Buddhist credentials" ("Definitions" 268). JoAnn Campbell would likely agree, since she maintains that any possible misappropriation of mindfulness traditions is made up for by how this usage can redress the larger problem of students' writing struggle (251). This view toward a pragmatic application of mindfulness is evident in Stephen Batchelor's *Secular Buddhism*, which argues that Buddhism needs to be a lived practice, a means to an end, rather than a protected belief system:

> the dharma is an expedient, a means to achieve an urgent task at hand, not an end in itself that is to be preserved at all costs. It emphasizes how one needs to draw upon whatever resources are available at a given time in order to accomplish what you have to do. It doesn't matter whether these resources are 'what the Buddha truly taught' or not.
>
> *(107)*

Batchelor cites a parable of a raft made of grass, a metaphor for the dharma, used to cross a river: once you've crossed, leave the raft behind. I tend to agree with Jeffrey Morgan: "It is likely that the Buddha would approve; the Buddha was ultimately concerned with the cessation of suffering, and not with metaphysical truths" (292). The suffering of writers and writing students is likewise real—and at times so acute as to determine students' academic and professional futures. I am inclined to think that the use of mindfulness in composition pedagogy would receive an approving nod from a double-chinned Buddha.

The Approach of This Book

In general, I avoid italicized terms from Pali or unnecessary jargon from primary Buddhist texts. My intent in this book lies primarily with writing instruction, writing students, and writing teachers, and to that end, I have pragmatically employed a range of ideas from Buddhism. Those more interested in Buddhist history and belief than in writing might be surprised by the buffet-style Buddhism offered up in this book, with its range of practices from Japan, China, Thailand, Vietnam, India, and the United States and its concoction of centuries and cultures. However, Buddhism is a highly amalgamated belief, not even called "Buddhism" as such by some of its practitioners, and even the most well-known sutra attributed to the Buddha were not written down for 100 years or more after his death; not to mention that by all accounts, the Buddha never wrote anything down (Lopez 12–13). Historically, Buddhism has been highly adaptive to its cultural and social context, modifying itself to accommodate a local culture (Shih 108). The theory of mindful writing covered in this book also doesn't require the lotus position, a zafu, tea ceremonies, vegetarianism, koan, incense, or chanting, although a little time spent on a meditation cushion, like time spent freewriting, probably wouldn't hurt anyone's practice. My emphasis in this book is on writing, and I don't pretend to be a Buddhist scholar, just an advocate of mindfulness for writing.

The sequence of chapter topics is intended to move readers from present awareness to a single rhetorical factor (intrapersonal rhetoric), back to another grade of present awareness (verbal emptiness) and a second rhetorical factor (affective responses to writing), replicating the clutch and release of mindful cognition. In Chapter 1: Present Moment, Writing Moment, I discuss how meta-awareness in composing is more efficacious with consideration of the Buddhist perspective on mindful metacognition. I explore the Satipaṭṭhāna Sutra or "The Foundations of Mindfulness," a sutra detailing the systematic awareness of actions, feelings, and thoughts with sufficient perspective to make necessary adjustments. The sutra shows us what to look for in a metacognitive practice for writing and presents a new set of rhetorical factors—impermanence, intrapersonal talk, and materiality—with a shaping influence on the writer, the writing act, and the writing outcome, though these factors are often ignored as we train students to essentially look over and above their immediate writing context. I turn to the conceptual metaphors of *writing-as-a-process* and *rhetorical situation* as vehicles for mindlessness and reexamine both through the lens of Buddhist present moment awareness.

Chapter 2: The Monkey Mind of Intrapersonal Rhetoric studies a topic which would likely attract readers who are interested in the discursive; namely, the near constant arising of inner discursivity that occurs in human consciousness as a factor in a rhetorical situation. Intrapersonal rhetoric is the most immediate discourse available to students: first on the rhetorical scene, its shaping influence on subsequent external rhetoric should not be underestimated. In addition to purveying content for writing, the intrapersonal foments preconceptions and other writing liabilities. Writing anything involves a private, internal discursivity that takes on the paradoxes of self and other, ones illuminated by Buddhist theories of no-self. Learning to control inner rhetoric through mindful metacognition benefits student writers in a myriad of ways in addition to finding content—not the least of which is the confirmation of writing ability and dissolution of writing blocks.

In Chapter 3: The Verbal Emptiness of Mindful Invention, I return to present-based awareness but introduce the complexity of verbal emptiness. I propose that mindful invention means facing the moment for the purposes of writing and that initially, the moment is nonverbal before intrapersonal content rushes in. The moment starts off as non-discursive and turns discursive. The Buddhist concept of the mutuality of form and formlessness is *sūnyatā*, an emptiness in which it is said that all things, not just the human ego, lack independent existence. *Sūnyatā* is the repudiation of a particular kind of existence (independent and permanent), replaced in a mindfulness perspective with an interconnected and continuously changing one. Examining the Buddhist *Heart Sutra*, renowned for its instruction on the nondualistic interplay of form and formlessness, I discuss ways to approach prewriting to engage verbal emptiness.

In Chapter 4: Mind Waves, Mind Weeds, Preconceptions, I adapt Suzuki's mind waves and mind weeds as a nondualist understanding of the affective dimension in writing, one that does not rigidly differentiate between logos and pathos, reason and emotion, as part of a Right Effort for composing. Mind waves and mind weeds are mental formations that momentarily disturb the emptiness of the mind without being separate from the mind. Essentially, the concepts of mind waves and mind weeds provide a spectrum of non-discursive to more discursive affective experiences based in the present. These experiences are elided from conventional theories of writing, deemed irrelevant to the work at hand, all the while exerting influence on students' writing. The Buddhist practices of *metta* or equanimity (calm, non-reactive mindset to internal developments) and *maitri* (calm, non-reactive mindset to external developments) are methods to encounter weeds and waves—and better manage a rhetorical situation.

In Chapter 5: Their Ability to Write Is Always Present: Establishing a Disciplinary Context for Mindfulness, I argue for the inclusion of present temporality in national writing standards and, relatedly, a more mindful approach to self and audience, completion and incompletion. Mindless learning is visible at the disciplinary level of writing instruction as reflected in national position statements by organizations of writing educators in the United States. In these policies, mindlessness manifests in future-oriented rhetorics and praxis; the diminishment of interiority and writer presence; the mishandling of the preverbal and formlessness; and the omission of the affective experience of writing students, not the least of which is the suffering caused by writing instruction. The impact of these policies of mindlessness manifests in classroom instruction on a school-by-school basis, affecting the next level of curricular and policy documents at colleges and universities. Mindlessness in writing policy can be redressed by implementing a Buddhist approach to metacognition and through process and rhetorical learning based on the present rhetorical moment.

The idea for this project began in 1992 with an item I packed when I moved to Iowa City a few weeks after my undergraduate commencement to study at the Iowa Writers' Workshop. The item was a thin paperback on Buddhist meditation that I'd removed from the shelf over the television in my parents' living room and which I read on my first night alone in the unknown town at the university hotel. The highly competitive environment of the Iowa Writers' Workshop at age twenty-two, compounded by a lack of instruction in the process of creativity and the uncertainty of a career as a poet, left me unable to write with ease for nearly a decade, although occasional breaks would happen in the block whenever mindfulness was at play. I'd often felt I was hostage to a massive problem. I received little guidance on the difficulties of writing, difficulties I later realized were directly attributable to my perceptions of the writing moment. Every now and then, I'd catch a glimpse of another writer's process. I'd overhear how a certain award-winning teacher practiced letting his work go fallow for a year after finishing a book or how another teacher obliquely mentioned a breathing technique to start new pieces. No one talked about ways to generate and continue writing or how to manage audience proximity. No one explained that what might look like a writing block could actually be a necessary delay or the natural functioning of the unconscious.

A few years later, mindfulness reemerged in my writing process as I was on the academic job market and writing a seminar paper for my last course in another graduate program, a class serendipitously with Peter Elbow. I worked with a Post-It stuck to the library computer at Smith College scrawled with either "Buddha" or "Present Moment," though I had no idea why. Another lengthy writing block ended with the severely premature birth of my first child after I blended mindfulness with writing as a way to cope with those difficult hospital weeks—and the block didn't return during my doctoral studies. In the years since, I have changed from an isolated writer with many notebooks of private writing and minimal and highly cramped drafts, someone who spent seasons at her desk with little to show for it, to who I am now, a writer who is at peace with her writing and publishes in several genres—scholarly, poetry, creative nonfiction. In 2015, a peak year to date, I published three scholarly articles, one review, two creative nonfiction essays, fifteen poems, and an edited pedagogy collection, followed in 2016 by a scholarly article, a book review, two creative nonfiction essays, and fifteen poems, in addition to conference presentations. And in 2017, mindful writing accompanies me every pre-dawn session of working on poems and then later in the day, scholarship or prose, workplace writing, or transactional personal writing. I am deeply indebted to the moment.

Moment by moment, a prolific bell is to be struck. Bells need to be rung again and again in order for sound to continue. All instruments must be repeatedly touched or blown through for creation to continue, but a bell, more than other instruments, appears to issue a single discrete note. Seeking sustained ability, few people would care to associate writing with a bell. A bell appears too fragmentary. A bell, however, resembles the process of mindful writing or that which allows our students to write prolifically, fluently, contentedly, and effectively. Trumpets announce and pianos make complex arguments, but the bell, the bell hinges us to time. A bell says *here we are, here you are.* Bells are demarcations of time—school bells, door chimes, alarm clocks, cell phones, church towers, gongs—that say: *This is over; this is starting. You are here; you are here now.* In each moment, the bell needs to be contacted again, or the sound will die. Our first-year writing students also need to make contact with the moment again and again to find the sounds of writing. Students need to look again and again into the contents of their intrapersonal, interpersonal, and intertextual selves in each passing moment to be able to write with ease. Listen attentively to the fluctuations of the bell and you will find yourself waiting for the moment when the bell is struck again: so, too, is it with writing. Watching the fluctuations of the rhetorical moment, we will find ourselves patiently receptive to the next phrase or idea, ones we can rest assured will pass by. During composing, countless moments pass before our inspection carrying phrases, concepts, images, and sensations like boxcars decorated with intriguing graffiti. What I am describing is mindful writing or the awareness of the discursive potential inside the panel of every moment. I am suggesting that our students' ability to write is always present. With training, reflection, and practice, we can all become prolific bells, and we can help our students become practitioners of this mindful composition.

Like a meditation bell, this book is meant as a constant call back to the present of writing in first-year composition.

1

PRESENT MOMENT, WRITING MOMENT

> Time goes from present to past.
>
> —*Dogen-zenji*

All writing, like any other activity, occurs during a present moment—not in the past, not last year, not in the future, not tomorrow. I have never written tomorrow. You have never written tomorrow. Despite the significant drawbacks and consequences to overlooking the present while writing, most people tend to do precisely that in a mindlessness that's inherent in human nature and reinforced by mainstream writing instruction. The majority of students bestow disproportionate consideration on the future and the past as they write, and those temporalities become overly influential as constraints on the rhetorical situation. Meanwhile, present-based rhetorical factors which could otherwise provide valuable insights about emerging content and writing activity are sacrificed in exchange for considerations of how writing carries future impact or is impacted by future factors. Much is lost for students with a misplaced present moment: it's a poor bargain, because students surrender rewarding composing experiences for stress, frustration, boredom, fear, and shortchanged invention. Writing becomes a markedly different experience if students think of it more consistently as part of a discrete now: every moment becomes an inventive moment due to the establishment of a calm, non-evaluative, and observant outlook that promotes receptivity to new ideas and critical thinking.

A student put in a situation without the moment is deprived of rhetorical information and is at a disadvantage: the student is asked to consider the hypothetical and function on assumptions while ignoring the information about content, process, and affect that is actually immediately at hand. The conundrum is that this forever future rhetorical date has many students preparing for impact, locked in a defensive rebuttal mode, anxiously reaching after already finished polished texts and comparing themselves with more skilled future-based and apparitional, nonexistent versions of themselves. What results is a disembodiment that has been implicated in a lack of perceived purpose, a silo-ing of lived experience from classroom instruction, and a failure to recognize and value one's own subject position (Berila 37; Kirsch and Ritchie; Wenger *Yoga Minds* 19). Elsewhere, I have joined others in calling for a more nuanced sense of the temporality of writing in composing pedagogies (Enos; Peary "The Role"; Short), including John Lofty, who in *Time to Write* described the disjuncture between students' lived sense of time as part of a close community and the instructional time of writing courses. Lofty identifies eight "timescapes" experienced by high-school students in this seafaring community in Maine, which in turn affected how students engage with learning—activity time, natural time, clock time, school time, social time, sacred time, existential time, and technological time. It is my contention that school time is most egregiously disconnecting when it removes students from their present time.

When was the last time you or your student wrote in the future? I may announce that I will be writing next Tuesday at 2 pm, but no writing, as procrastinators cringe to know, is happening as I make my pledge. The only thing that is actually happening *is* my pledge; when 2 pm next Tuesday rolls around, who knows where circumstances will find me? The same goes for the past: we don't produce new writing in our past. As Thich Nhat Hanh says, "We have an appointment with life in the present moment. If we miss the present moment, we miss our appointment with life" (*The Path* 29). Or, as Janis Joplin crooned, "Tomorrow never comes." In composing pedagogy, the temporality of writing generally leans toward the future perfect tense, framing writing as a past event occurring in the future (i.e.: "I will have completed the final version by 5 pm"), and this tendency is related to repression. Anxious, the person writing tells herself promises to ward off "anticipated rather than experienced sensations" for the future (Short 369). Kathleen Blake Yancey's baseline definition of reflection, that shoe-in for metacognition, skirts the present moment by jumping to a future moment with reflection as "*looking forward* to goals we might attain" and then jumping to a past time by "*casting backward* to see where we have been" (6). Yancey corrects this position by subsequently distinguishing between types of rhetorical reflection: "reflection-in-action" involves a writer's activities of projection and retrospection as they occur in an ongoing and immediate moment; "constructive reflection" is the accumulation of process insights and agency over the course of multiple moments and multiple documents (13–14). Because there isn't a single text which wasn't made in the present, by extension, there isn't a rhetorical situation that isn't housed in the moment or a writing process that isn't momentary. Regardless of whether we're cognizant of it, the present moment is a factor in every single composing situation. *This point alone should be sufficient cause to think carefully about the role of the present in the writing situation—to essentially rethink our praxis of metacognition in composition studies.* An opportunity for this rethinking is already built into the pedagogy and theory of the field; it takes the form of the discipline's interest, evinced in national writing policies and standards, in building students' metacognition through writing. I agree with Christy I. Wenger, who sees a role for mindfulness implied in the learning outcomes posited by the *Framework*, for, as she says, these guidelines don't simply emphasize the development of rhetorical knowledge but, rather, pair rhetorical training with training in awareness. Wenger concludes that "we might see the Framework as underscoring the importance of developed writerly awareness, or of approaching writing mindfully" (*Yoga* 104). The situation is promising, since elements of mindfulness already apparent in established composing praxis.

In Buddhism, the present serves as a sort of Petri dish from which impermanence can be detected along with signs of attachment, the triggers of suffering. The Buddha's explication of the Four Noble Truths doesn't reference "the present" or "mindfulness" per se. Both terms rose to prominence relatively recently in contemplative practices (Gethin *Foundations*). Jessie Sun has pointed out how the term "mindfulness" had a long history outside of Buddhism, and surfaced in fourteenth-century Christian practices; it only became associated with Eastern thought at the late date of 1881 with the translation of Pali scripture by T. W. Rhys David. To put this in perspective, "[e]ven the term *Buddhism* is of recent vintage" and practitioners today in Sri Lanka think of it as "sāsana," translated minimally as "the teaching" (Lopez 11–12). Instead of mindfulness or the present moment, the Buddha mostly referred to transience, impermanence, clinging, and dependent conditionality: observation of the present moment worked as the method to grapple with the causes of human suffering. The Buddha's foundational treatise is summarized in the third of the Four Noble Truths: "whosoever of the monks or priests regards the delightful and pleasurable things in the world as impermanent (*anicca*), miserable (*dukkha*) and without an Ego (*an-atta*), as a disease and sorrow, it is he who overcomes the craving" (Goddard 31). In diagnosing human experience of suffering, the Buddha referred to the impermanence of phenomena and how people suffer because they try to retain certain pleasures, warding off inevitable fading, deterioration, decline, diminishment. Our collective tendency to resist transience is evident when we attempt to repeat pleasant experiences (a fine meal at a restaurant, praise from one's boss, a souvenir from a Mediterranean vacation).

Immediately after providing the four central tenets of his approach, the Buddha delineated the steps to be taken to enact his theory, measures known as the Eightfold Path. Right *Mindedness* appears as the second of eight follow-up practices to eliminate suffering (Right Understanding, Right Mindedness, Right Speech, Right Action, Right Living, Right Effort, Right Attentiveness, and Right Concentration). Right Mindedness dealt with perceiving thoughts arising during the moment through Mundane Right Mindedness (noticing thoughts of lust, ill-will, cruelty, and so forth) or Ultramundane Right Mindedness (noticing more abstract "Verbal Operations" such as thinking and reason) (Goddard 42). For the purposes of writing, the present moment can be observed to determine a writer's dynamic with impermanence, as both a content provider and a source of mind waves and mind weeds, to alleviate writing-related suffering.

In this chapter, I explain how meta-awareness in composing is more efficacious with consideration of the Buddhist perspective on mindful metacognition. The general principles of a mindfulness pedagogy for composition are the perception and exploration of the fleeting, impermanent nature of all phenomena for the purposes of learning and writing and the acceptance of mental formations (including emotions and the registering of physical sensations) without sorting or attachment—all occurring in real time. Metacognition seems to be built into the very notion of mindfulness, since the translation of the Pali word for mindfulness, *sati*, is a "calling-to-mind, being-aware-of certain specified facts," essentially an invitation to "remember to remember" (Rhys David qtd. in Gethin "Definitions" 265). In contemplative pedagogies, mindfulness is valued as an educational practice for fostering metacognition because it's "more than just simply being aware. It involves being aware of aspects of the mind itself" (Siegel 5). Buddhist practices lead to metacognition and critical thinking because they "train awareness of awareness" such that "meta-awareness takes the form of witnessing thoughts, emotions, and sensations as they arise from moment to moment, and observing their qualities" (Thompson 52). Across traditions of Buddhist theory, several terms designate metacognition: *clear seeing or comprehension*, *bare attention*, *luminous mind*, *bodhichitta*, *Buddha Nature*, and *beginner's mind*. Each hones metacognitive skills by promoting receptivity to newness and alternative perspectives after setting up a birds-eye view of one's consciousness. The variants underscore the importance of seeing our thoughts as they are, noticing the ways in which we heap evaluations or categories onto our thoughts, leading to the installment of preconceived notions. A main point I seek to make is that greater awareness of present temporality yields a whole other set of rhetorical factors, ones often overlooked as we train students to essentially look over and above their immediate writing context, and that these are the basis of a mindful metacognition that can assist our curricular goals for learning and writing.

In the first section, I overview Buddhist awareness of the present as a groundbreaking set of teachings on meta-awareness as evinced in the Satipaṭṭhāna Sutra or "The Foundations of Mindfulness." This sutra offers a systematic way of developing awareness of actions, feelings, and thoughts with sufficient perspective to make necessary adjustments. The sutra shows us what to look for in a metacognitive practice for writing and presents a new set of rhetorical factors—impermanence, intrapersonal talk, and materiality—which have a shaping influence on the writer, the writing act, and the writing outcome. In the second section, I turn to the conceptual metaphors of *writing-as-a-process* and *rhetorical situation*, which undoubtedly exert as large-scale an influence on the field of composition as national position statements (certainly one that's lasted longer) but have done so while perpetuating mindlessness. In the final section, I reexamine those constructs of writing process and rhetorical situation through the lens of Buddhist present moment awareness and the trifecta of rhetorical factors as integral to a more ontological and epistemological writing experience. With this meta-awareness, writing happens without a past, without an illusionary future, but within a deeply inventive present moment characterized by calm, low-stakes exploration and privacy, even during later-stage composing phases. In this approach of mindful invention, every moment is a present moment, and therefore every moment can be a writing moment.

The Sutra of Sutra: An Early Praxis of Metacognition

The teaching which encapsulates present metacognition instruction is the Satipaṭṭhāna Sutra or "The Foundations of Mindfulness," offered circa 544 BC at the First Council of Monks near Delhi. It's "generally regarded as the most important Sutta in the entire Pali Canon" (Walshe 588). This sutra was said to provide the most direct path to enlightenment; its content leads directly into the most renowned of Buddhist concepts, that of the Four Noble Truths. In "The Foundations of Mindfulness," the Buddha described the development of *citta* or consciousness, a metacognition or a thinking about thinking, a knowing about knowing, which leads to an ability to step back from thought, to progress from noticing thought to then noticing its impermanence, and then developing an outlook of detachment and bare attention. In doing so, the Buddha offered a curriculum of critical thinking that countered an essentialism fashionable in his day in which false views of an autonomous self were pronounced. By claiming that "phenomena are conditioned and changing," the Buddha refuted the popular notion, not dissimilar to Platonic universal truth, of an "essentialist view of self" that enabled the oppressive caste system in Northern India in the sixth century BC (Maitra 363; Stanley 69). Static notions of self (or of other matters) risk incurring faulty perceptions, missed connections, and limited information, as does any assumption.

The Satipaṭṭhāna Sutra culminates in a one-paragraph demonstration of how its concepts apply to the Four Noble Truths; in a longer version, the Mahāsatipaṭṭhāna Sutra or "The Greater Discourse on the Foundations of Mindfulness," the Four Noble Truths in their entirety close the document. It was typical of the Buddha's teaching style to offer a range of options to reach enlightenment; depending on the learner, the Buddha adjusted his pedagogy in what Donald S. Lopez calls his "skillful means" such that "each person hears what is most appropriate for him or her, spoken in their own language, and each person in the vast audience thinks that they are receiving private instruction from the Buddha" (69). In his periphrastic practice, the Buddha instructed crowds of laypeople or householders, *bhikkus* and *bodhisattva*, or individuals who have committed themselves to Buddhist practice but have delayed enlightenment, and *arhat*, or those who have cleared away all obstacles to their enlightenment. Jienshen F. Shih describes the Buddha as "a skillful and creative adult educator" who used "different ways of interaction between the teacher and the disciples" (110). So the sutra closes with a count-down of the amount of practice required to reach enlightenment through its direct method, starting with a practice time of seven years, moving to seven months, and then reducing itself to just seven days, because "one of two fruits could be expected for [the practitioner]: either final knowledge here and now, or if there is a trace of clinging left, non-return" (Ñāṇamolí and Bodhi 155). None of the Buddha's teachings were recorded during his lifetime, and the Satipaṭṭhāna Sutra is recounted by Ānanda, a blood relative, personal assistant, and close disciple of the Buddha reputedly possessing a prodigious memory—Lopez says Ānanda is frequently the "I" in the "Thus I have heard" signature opening phrase of many sutra (108).

The Satipaṭṭhāna Sutra demonstrates how to establish a non-reactive observational stance and notice impermanence—a great pairing that leads to a metacognition of the present. In his close study of this sutra, Anālayo has described it as instilling a "protective mindfulness" in which the "task is not to avoid seeing things altogether, but to see them without unwholesome reactions" (24). Noticing impermanence has been described as the most radical alteration in individual thinking resulting from this method: "the first powerful impact on the observer's mind will probably be the direct confrontation with the ever-present fact of Change" (Thera 36). The Buddha explains the four foundations of mindfulness as contemplation of body, feelings, mind, and mind-objects; metacognition enters because what's advocated for each category is an awareness of itself: "the body as body," "feelings as feelings," "mind as mind," and "mind-objects as mind-objects" (Ñāṇamolí and Bodhi 145). The verb phrase preceding each of these categories, "a bhikkhu *abides contemplating*" (emphasis added), captures the essence of mindful activity—remaining (not departing into distraction) in order to reflect. This activity sets up the Four Noble Truths at the end of the sutra because it steers practitioners toward actuality or the Dharma by helping them gain control in particular over mind-objects or mental formations (Ñāṇamolí and Bodhi 1194): "Here a bhikkhu understands as it actually is: 'This is suffering'; he understands as it actually is: 'This is the origin of suffering'; he understands as it actually is: 'This is the cessation of suffering'; he understands as it actually is: 'This is the way leading to the cessation of suffering'" (Ñāṇamolí and Bodhi 1194; 154).

The goal behind all four foundations of mindfulness is to further metacognitive insight, which ranges from simple awareness to awareness of transience to detachment. Bare awareness occurs with the realization of body, feeling, or mental formation as light and detached, so that the student "abides independent, not clinging to anything in the world" (Ñāṇamolí and Bodhi 146). For each of the four areas, practitioners are provided with a theory of awareness as well as exercises, in line with the Buddha's three-part pedagogy of study, reflection, and practice. One exercise entailed contemplating corpses "thrown aside in a charnel ground, one two, or three days dead, bloated, livid, and oozing matter" or different skeletal states, "fleshless" yet "smeared with blood" or "disconnected bones scattered in all directions" (Ñāṇamolí and Bodhi 148–149; Shih 103). Another list catalogs feelings including "delusion … contracted mind … distracted mind … exalted mind … surpassed mind … concentrated mind … unconcentrated mind … liberated mind … unliberated mind" (Ñāṇamolí and Bodhi 150). Mental formations are categorized as Mundane (any emotion) or Ultramundane (abstractions less tied to worldly interaction, including "Verbal Operations" and "whatsoever there is of thinking, considering, reasoning, thought, ratiocination, application") (Goddard 42). "Such arms-length distance" when incorporated into a twenty-first-century classroom, says Tobin Hart, allows students to perceive critically what might otherwise be assumed, to detach from ideas and gain perspective rather than becoming knee-jerk reactive to thought: "Being aware of the content of our consciousness is not only an important element in emotional maturity but also a marker of deepened cognitive functioning … Self-observation and reflection help to expose and deconstruct positions of role, belief, culture, and so forth to see more deeply or from multiple perspectives" (33). This metacognitive effort is significant because of the ubiquity of mental formations whereby "we are always giving our attention to something. Our attention may be 'appropriate' (*yoniso manaskara*), as when we dwell fully in the present moment, or inappropriate (*ayoniso manaskara*), as when we are attentive to something that takes us away from being here and now" (Hanh *The Heart* 64). Two cognitive benefits ensue from avoiding what the Buddha called "a thicket of views" and adopting a more detached view on phenomena, including the self: first, we are better able to perceive how phenomena are in constant flux (which can lead to rich improvisational experiences and recognition of multiple perspectives); second, we are less likely to be tricked into elaborate storylines (narrated by our intrapersonal babble), which make us more reactive than thoughtful.

The perception of an ever-shifting actuality in the Satipaṭṭhāna Sutra alters our sense of other temporalities and requires a specific disposition toward uncertainty—adding appreciation of contingency to the qualities of metacognition. In Buddhist philosophy, "Form is transient, feeling is transient, perception is transient, mental formations are transient, consciousness is transient" (Goddard 27). A refrain in the Satipaṭṭhāna Sutra underscores the perception of transience of all entities: the practitioner "abides independent, not clinging to anything in the world" (Ñāṇamolí and Bodhi 146). No moment completely resembles the next due to this continuous changeover. When the present serves as the chief epistemological and ontological platform, talk about the past or the future becomes a discussion of mirages. While we often host future-oriented thoughts in this present moment, we never experience the future first-hand in the way we do the present. It is always a fraction of the present, a subsidiary, always subsumed by the present, always the content and never the vehicle. The future is second-hand experience, and so is the past. Yet the human tendency is to think of the future and to plot, "'I must do something this afternoon,' but actually there is no 'this afternoon.' We do things one after the other" (Suzuki 30). Does writing (or anything) have a future? A radically groundless view would answer in the negative. Writing occurs without a past; writing occurs without a future. Buddhist approaches to present temporality are aligned with presentism, a metaphysical outlook that "claims only present objects exist," in contrast to the growing block view, which supports the reality of the present and past but not the future, and eternalism, which places all classifications of time "on the same footing" (Meyer 87). To be able to remain mindful and extend perception of the present, practitioners need to develop a non-judgmental acceptance of the changing moment, welcoming rather than resisting contingency; these are metacognitive functions mentioned early in the "The Foundations of Mindfulness" sutra.

While noticing the present means attending to changes arising in our momentary context, filtered through our perceptions, it also means attending to changes in our intrapersonal rhetoric, that interior river of phrases, images, shifts in voice, imagined conversations with imaginary interlocutors and future audiences. Sitting on a cushion, a meditation practitioner might be aware of that inner talk for the first time in hours. Previously, self-talk had manipulated the individual, steering their hours, formulating their outlook, interactions, and movements, all without the purview of the person. The person functioned as a sort of sleepwalker, obeying the commands of self-talk, copying its moods, parroting its opinions almost mechanically. When in a reactive mindset, we take the bait of inner dialog and allow ourselves to be dragged away from present circumstances into a hypothetical circumstance. Bhante Gunaratana describes how "[n]ormal conscious thought ... is ponderous, commanding, and compulsive. It sucks you in and grabs control of consciousness. By its very nature it is obsessional, and it leads straight to the next thought in the chain, with apparently no gap between them" (70). In contrast, discernment of inner talk is indicative of the person's critical perspective on internal self-rhetoric. Gunaratana continues, "A thought you are simply aware of with bare attention feels light in texture; there is a sense of distance between that thought and the awareness viewing it. It arises lightly like a bubble, and it passes away without necessarily giving rise to the next thought in that chain" (70). As I will discuss, mindful writing means replacing knee-jerk mental habits of judging, categorizing, or editing with the examination of self-talk for the purposes of generating content and shaping one's self-ethos as a writer.

Two Vehicles for Mindlessness in Composition Pedagogy

As one of two prevailing tropes in composition pedagogy, the conceptual metaphor of *writing as a process* transmits assumptions about writing which can be detrimental to metacognition. Conceptual metaphors structure everyday functioning by organizing our perceptions and actions, and they incur conglomerates of thought as "metaphor expressions [that] recruit larger metaphoric concepts" (Eubanks 46; Lakoff and Johnson). *Writing is a process* is an ontological metaphor, a specific category of conceptual metaphor that helps us "comprehend events, actions, activities, and states" (Lakoff and Johnson 30). If we say something is a process, this means that our view of things entails a series of steps and a posited end point (the details about which are unspecified). A process normally takes us somewhere. For instance, if I say that obtaining a passport is a process, I might mean paperwork to be completed, a trip to the post office for an acceptable photo, money to be sent to a government agency, and a period of waiting—implying a regularity of experience and an expectation that the different moments will culminate in an outcome. As a conceptual metaphor, *writing is a process* may shape a writer's actions by providing comfort in reducing the unknown (there's an end step to reach) and suggesting that the writer's movements are regularized and regulated. In actuality, there's likely little that's similar between the real, moment-to-moment activities of a writer and the mirage cast by the metaphor.

Moreover, lacking a truly open, moment-by-moment awareness, the metaphor is actually weighted toward outcome, skewing composition pedagogies tacitly toward written product. To work on metacognition means to develop a mental practice of perceiving impermanence, since to be aware of one's thoughts necessarily means seeing them in flux. Otherwise, we become locked into static views through clinging and attachment or are carried off by seductive storylines in our minds. Daydreams of written products which are yet to be completed are precisely that—daydreams. James Seitz points out that "Even process pedagogies ... almost always want the process, however it begins, to lead finally to fully integrated writing, complete with orderliness, transitional devices, and 'development of ideas' necessary for the reader's untroubled consumption of form" (820). As Ellen Langer reminds us, "an outcome orientation in social situations can induce mindlessness. If we think we know how to handle a situation, we don't feel we need to pay attention" (*Mindfulness* 34). For writing students, the presumption that a progression of phases will culminate in an end point is more subtly consequential than any lockstep through prewriting-writing-drafting-finalizing. This product bias hidden in the metaphor subtly sands away at other learning goals and habits of mind sought in composition instruction.

A clear example of the complexities of the process metaphor is evident in Donald Murray's seminal 1972 "Teach Writing as a Process Not Product," in which he suggested that instructors avoid a static view of student composition and instead treat it as occupying time and dependent on factors such as social interaction, life experience, and individual work pace. He famously contrasted static "dead" writing with "live" writing: "Most of us are trained as English teachers by studying a product: writing. Our critical skills are honed by examining literature, which is finished writing ... And then, fully trained in the autopsy, we go out and are assigned to teach our students to write, to make language live" (1). However, sequencing and measuring are writ large in Murray's approach, more of a sense of *chronos* than of *kairos*, and the timing of writing is treated as a quantifiable resource that a student uses in a procedure with already-identified stages: 85% for prewriting, 1% for the first draft, and 14% for rewriting, if you do the math (2–3). Thus, the actions Murray prescribed for instructor and student reflect the message of regulation in the conceptual metaphor of writing as a process. Murray attributed Sondra Perl's work with felt sense to helping him to "see an instantaneous moving back and forth during the writing process" ("The Essential" 10). In later work, Murray came closer to present awareness—particularly in his explorations of prewriting and felt sense.

Similarly, the conceptual metaphor of the rhetorical situation poses a problem of mindlessness in writing praxis and diminished meta-awareness. Lloyd Bitzer's equally influential 1968 construct of the rhetorical situation unpacked the circumstances behind composing—audience, exigence, and constraint. Bitzer defined the rhetorical situation as a "natural context of persons, events, objects, relations, and an exigence which strongly invites utterance; this invited utterance participates naturally in the situation, is in many ways necessary to the completion of situational activity" (4). Although commendable for drawing students' attention to the components of writing, the construct falls short of placing the act of writing in time. It's the difference between noticing an ink bottle on the desk (staying aware enough to see an object in one's immediate surroundings) and noticing that one's noticing of the ink bottle is occurring in real time (metacognition)—the difference between seeing or creating a detail and being aware of the thinking that leads up to the detail's creation—or, in the case of mindful writing, latching onto one's metacognitive awareness to find new material for writing. The static, atemporal quality of Bitzer's construct is visible in the verbs in Bitzer's definition—rhetorical factors are personified (the writer is nowhere in sight) and writing becomes a pair of nouns, "completion" and "activity." The atemporal is further evident in how he defines a situation as a combination of circumstance and structure (1). As what Lakoff and Johnson call a container type of conceptual metaphor, the construct that writing is a situation suggests drama, a stage setting, almost like a one-act play: who are the actors? What's going on? What's at stake? The metaphor lends the idea of an occasion in which something is unresolved or unaccomplished, a tension, in which the student becomes an arbitrator needing to diffuse an interpersonal conflict. Metaphoric drama is found in the comparison that opens Bitzer's "The Rhetorical Situation": "If someone says, That is a dangerous situation, his words suggest the presence of events, persons, or objects which threaten him, someone else, or something of value" (1). Just as *writing is a process* cloaks certain real-time actualities, *writing is a situation* conceals the cognitive and embodied changes undergone by the student while writing. Furthermore, attention is diverted from the choices and rhetorical moves made by the writer to the directed movements of the players, those rhetorical factors, on the stage of the situation.

Subsequent interpretations of the rhetorical situation attempt to avoid a static account of writing by adding a modicum of impermanence and cutting out the idea of preexisting material. By including how rhetorical factors might change during the rhetorical situation, scholars have argued against a fixed sense of the rhetorical situation (Biesecker; Phelps; Reynolds). Jenny Edbauer states that "we might also say that the rhetorical situation is better conceptualized as a mix of processes and encounters; it should become a verb, rather than a fixed noun or *situs*" (13). In an early response to Bitzer, Richard E. Vatz argued that rhetorical factors are not autonomous, preexistent elements but, rather, established by students during the writing moment (157). Scott Consigny offered a compromise between Bitzer and Vatz that allowed for both prefabricated and created elements to a rhetorical situation. According to Consigny, a rhetor's "integrity" involved making use of established strategies for any rhetorical situation, and a rhetor's "receptivity" involved greater flexibility, tinkering with strategy based on the individual rhetorical situation. All of this is important to consider in light of mindful learning, since, as Benzion Chanowitz and Ellen Langer say, "Mindful involvement continues in the recognition that the environment is constantly shifting as one becomes more involved with it ... [and] that the environment is constantly offering new opportunities for grasping surer control of the material" (112). Moreover, this material isn't merely passively discovered by the learner but is instead, as Vatz claims, produced or evoked by him or her (Chanowitz and Langer 113). With flux as a factor, these adaptations of the metaphor of rhetorical situation facilitate greater metacognition in composing in ways corresponding to the Buddha's Satipaṭṭhāna Sutra.

Present Temporality, Process, and Situation

By adding present temporality to the metaphor *writing is a process*, we reconceptualize process as a movement in time that entails watching phenomena as they arise and fade with a detached, clear outlook. The practice of writing, contained in a process and movement, becomes equivalent to the steps taken by the practitioner in the Satipaṭṭhāna Sutra who heads to the base of a tree or an empty hut in the forest, sits in a meditation posture, and "establishes mindfulness in front of him" (Ñāṇamolí and Bodhi 145). The entire composing act becomes metacognitive; its guiding principle is to remain aware of one's thoughts, feelings, and actions and consequentially be more present in one's writing. Composing thus resembles the ontology that Robert Yagelski seeks for writers, in which "writing is a way of being in the world" such that "[w]hen we write, we enact a sense of ourselves as beings in the world" (3). This meta-aware process helps students perceive content for drafts from within that changing array of phenomena. Students also locate assumptions they may be entertaining about where they are in the development of their pieces. The significance of the first capacity is that it transforms any moment in the writing project into an inventive moment—even seemingly tail-end stages of editing and proofreading may generate content. The second affordance of a metacognitive take on process, the ability for students to perceive their assumptions about where their piece stands in its development, is important because writing is a "psycho-physical" movement "more or less coordinated by ideas about what we are and should be doing" (Hjortshoj 11). Stymied student writers maintain rigid, preconceived notions about their actions in a writing process, as Mike Rose pointed out, and they may be caught in a schism whereby they *believe* they are in a certain phase of writing, but their writing behavior indicates something else. This lack of awareness happens, for instance, if a student thinks she is engaged in composing an early draft but is actually occupied with grammar or editing tasks.

For the purposes of composition, replacing the macro beats of phases with the micro beats of the momentary helps students make fewer assumptions about their own activities within a writing process. With mindful invention, conceptions of the timing of writing are not of static phases but instead whatever fills each arising moment, to be reset a few seconds later, cleared away by the windshield wiper action of the breath. In contrast, labels like "drafting" and "rewriting" categorize our actions and experiences, and as categories, they risk mindlessness and a diminishment of metacognition. A common trap is to rely on long-established categories without noticing how they shape perception and obstruct fresh perceptions of the moment, and this includes categories about the writing experience (Langer *Mindfulness* 11–12). While a standard recursivity model deals in blocks of experience, in a composing model based on the momentary, each "mind moment," to borrow Evan Thompson's term, potentially contains in rapid-fire several so-called phases of writing. Keith Hjortshoj's loops and spiral model for composing, for instance, posits "macro loops" and "micro loops"; the failure to notice macro loops might trap a student in the micro loops of compulsively working at the sentence or word level, all the while remaining oblivious as to what they're actually doing during that moment of writing (33). Hjortshoj points out that struggling writers often make the mistake of expecting a writing process to be as sequential and linear as the experience of reading a completed text (27). Into this model, mindful invention incorporates an even quicker-changing, more detailed perspective based on the writing moment.

The instant we begin to draw air into our nostrils, we make contact with formlessness, a state akin to prewriting, and then we often hear intrapersonal rhetoric in the form of a phrase, half a sentence, an answer to our inner Q & A. By the time we reach a conscious exhalation, really a matter of a few seconds, we might entertain shaping or revision thoughts characteristic of an orientation toward interpersonal rhetoric. The whole gesture of writing is evident in any given moment. It's repeated with the next breath or unit of consciousness in an ultimate recursivity, since the student is returned to a beginner's mind, an outlook that drops assumptions built on mastery in favor of open acceptance (Fontaine 208; Suzuki *Zen* 21). Future-based stressors of premature audience consideration, anticipation of criticism, and unreasonable expectations of outcome are removed by the moment. They are replaced by a calm focus that is perceptive and receptive to possibilities, among other qualities prized for effective invention. Readers, grades, persuasive purposes, transactions, genre, issues of organization—these are important but secondary to the primary effort of accepting the moment in order to write.

This present temporality lessens outcome orientation during composing and, not unrelatedly, evaluative tendencies, both of which are replaced by bare attention on the composing moment. The normal goal lines of writing and instruction are no longer as dominant: a focus on the ever-changing moment pushes past final drafts, grades, and semester ends, potentially guiding writing into the long term if a student returns to build upon previous work. For example, if I maintain focus on the shifting experience of a certain text, I may be more likely to keep revising it, if that's what the moment calls for, than if my sense of the project is entirely determined by external outcomes of a deadline or publication. Moreover, since focusing on the present commands substantial effort, writers have less time to court or be taunted by self-generated evaluative thinking. Tracking the frequent changes of the present does not afford much time for evaluation, and this modifies how we usually sort experience into categories of positive, negative, and neutral—*this chair is uncomfortable; I like the verb she just used; that part of the methodology section bores me.* Most of us jockey to look good in the eyes of others and try to arrange matters for our comfort and security. As Chögyam Trungpa commented, "There is a problem with one's basic life, one's basic being. This problem is that we are involved in a continual struggle to survive, to maintain our position. We are continually trying to grasp on to some solid image of ourselves. And then we have to defend that particular fixed conception" (*The Essential* 76). Of course, evaluative thinking is connected to an outcome orientation: writers screen defensively for optimal content and writing moves that will lead to an envisioned outcome, usually one with an audience in mind. In doing so, a person fixates on an aspect that becomes static and disconnected from the interconnected mass of arising thought. The net result is that people's writing resembles one giant rebuttal, no matter what the genre or the stage of development of the text, as writers mix creating with editing in an attempt to win approval or admiration from future audiences. On the other hand, present-oriented writing remains mindful, in that it is a "non-analytic investigation of *ongoing* experience" as well as a "deliberate, open-hearted awareness of *moment-to-moment* perceptible experience" (Grossman and Van Dam 221).

Something is always arising due to ongoing impermanence and the radical contingency of the writing moment. Fragments, phrases, full sentences appear in the flotsam of consciousness. The Satipaṭṭhāna Sutra provides guidance on how to develop a metacognitive stance that incorporates a full range of arising formations—ideas and feelings as well as the body. Thus, the shifting present of writing summons awareness of physical sensations related to writing embodiment (posture, body movements of typing, sensation of jewelry or shoes on the feet, the feeling of the ballpoint pen held by a trio of fingers); awareness of the scene of writing (room or landscape, time of day, season, household sounds, drip-drop of melting snow); awareness of intertextuality (words in hard copy books, digital texts, on phones, in recalled conversations); as well as awareness of affective formations (alertness, discouragement, confidence). The sheer bulk of transient material encourages students to accept flaws, dullness, repetition, lack of clarity, and cliché and gives practice in low-stakes, informal writing. Because mental formations are transient as well as continuously arising, the material must to some extent be disposable writing: a writer simply can't hold on to it all, and this provides practice in detachment. Furthermore, concerns a student might feel about potential errors are alleviated by awareness that content is fleeting; wait a few seconds and something else will arise.

Intrapersonal Rhetoric and Materiality

Turning now to the metaphor of *writing as a situation*, two rhetorical factors step forward with greater attention to the present moment, specifically intrapersonal rhetoric and materiality, both of which complement conventional rhetorical factors of exigence and audience. As a writer pays increasing attention to present and decreasing attention to hypothetical, future-based occurrences, the mental formations of inner talk and sensations from embodiment become more noticeable and become areas for examination, as in the Satipaṭṭhāna Sutra. These twin factors are recognizable components which stimulate, sculpt, and possibly stunt discourse while at the same time remaining part of flux. They exemplify the way in which "consciousness arises dependent on conditions, but it also has its own causal influence on things" (Ñāṇamolí and Bodhi 350–351; Thompson 24). A mindful approach to composition necessitates monitoring of Mundane formations and Ultramundane formations, especially Verbal Operations, as part of students' practice with metacognitive awareness.

Intrapersonal rhetoric consists of whatever discursively arises as a mental formation during the moment. It is the language of the moment—any language, including single words, phrases, fragments, questions, recalled talk or reading, judgments about writing, descriptions of sensations, and crystallizations of past writing performances or imaginary dialog pertaining to anticipated future audiences. Intrapersonal rhetoric is the most immediate discourse available to students: it's first on the rhetorical scene, preceding interpersonal texts, and it's a requirement for finished writing: "*all* writing as authoring must be some revision of inner speech for a purpose and an audience" (Moffett "Writing, Inner" 233). (Intrapersonal rhetoric is so important to mindful composing that I devote Chapter 2 to it.) The intrapersonal is not necessarily recorded language: it doesn't always make it to the page or screen but arises and circulates in students' consciousness as they write, organizing and altering other constituents of the rhetorical situation. In fact, its frequent invisibility enables the intrapersonal to remain unnoticed and therefore invites mindlessness. In *Internal Rhetorics: Toward a History and Theory of Self-Persuasion,* Jean Nienkamp emphasizes the importance of internal rhetoric, since "talking to oneself is at least as ubiquitous and consequential as the various social language practices analyzed from a rhetorical perspective" (ix). This view is shared by Buddhists, who view the human mind's predilection for language production as both non-stop and widely divergent and picturesquely refer to this state as "a wild elephant and a drunken monkey" (Lopez 49). Nienkamp points to the persuasive nature of self-talk—how it influences moods and actions on a regular basis and flies under the radar of most people's consciousness—including while writing.

As a rhetorical factor, the intrapersonal is self-influential, leading to preconceptions, assumptions, attitudinal, and affective responses during writing activity. Frequently, students uncritically react to the intrapersonal, whereas a present-based metacognition can help them notice the appeals of intrapersonal rhetoric and utilize the intrapersonal as content for pieces of writing. According to Nienkamp, individuals possess rhetorical selves, which are the "ongoing product of individual acts of internal rhetoric," and because thought so deeply shapes our minds, it is rhetorical (x). A sub-category of internal rhetoric, "primary internal rhetoric," results when the unconscious persuades us, typically through an internalization of societal language (Nienkamp 113). Intrapersonal rhetoric thus navigates around nonverbal elements of consciousness—and sometimes responds to them—including nonverbal imagery, nonverbal physical sensations, and the blanks from the unconscious, that "gappy" discontinuity or the material that is "unavailable from the start," unlike "mindless ideas [which] were once potentially accessible for mindful processing" (Langer *Mindfulness* 26; Thompson 58).

Intrapersonal rhetoric is comprised of intertextual, socially inflected instances of language that nevertheless occur in solitude as the student writes in a particular present moment separate from a possible eventual reader. The material of the intrapersonal—its bits of voice, phrases, sentences—is intertextually informed by the language of others. At the point of utterance, the actual address of the intrapersonal is directed to no one but the self; that is, the actual or present audience of the rhetorical situation is the self. Intrapersonal rhetoric pirates non-stop from prior occasions of literacy—those "citations, quotations, allusions, borrowings, adaptations, appropriations, parody, pastiche, imitation, and the like" that connect one text to another text, one writer to another writer (D'Angelo 33). As Kristie Fleckenstein points out, this "discontinuous stream of internal monologues" transports culture to the mind through discursive codes ("Writing Bodies" 289–290). Many phrases arrive between invisible quote marks—bits of catchy song, recalled conversation, a rhetorical twist in a scholar's argument, or a memorable phrase from a book. However, intrapersonal rhetoric becomes diversionary when self-talk mutates into a hypothetical conversation happening in a hypothetical situation in which the writer is directly and immediately communicating with an imaginary audience. At stake is the way in which this rhetorical factor can lead us to habitually behave as though we were conversing in the moment of writing with another person or group, constructing the stage set of a hypothetical situation with imaginary occupants and assumptions as rhetorical factors.

Metacognition of the intrapersonal for composition does not lead to a total erasure of verbal formations, since the intent is to critically examine the intrapersonal for messages about the writing process as well as for content. In the Eightfold Noble Path, the Buddha explained Right Concentration as a discarding of an increasing number of mental formations, starting with the five hindrances or Mundane formations of "Lust, Ill-will, Torpor and Dullness, Restlessness and Mental Worry, Doubts," moving on to free oneself of the Ultramundane formation of "Verbal Thought"—our intrapersonal rhetoric—followed by a dropping of Rapture and Happiness, leaving the practitioner able to concentrate on the arising and falling of formations (Goddard 55–56). Similarly, in the Satipaṭṭhāna Sutra, a practitioner watches the moment as various mental formations arise, using their emergence to practice relinquishing attachments. Composition students, on the other hand, can be trained to linger on arising intrapersonal messages in order to analyze them for writing insights. It's also the case that recording the intrapersonal onto the page can be a satisfying and restorative experience for writers. Honoring one's intrapersonal rhetoric by making it visible and epigenetic can improve frustrated interpersonal communication, since ruptures in one's self-communication lead to ruptures in one's communication with others (Rogers). The usual time lapse between the different pace of our thoughts and our transcription of them through handwriting or typing can install an attitudinal constraint of calm by putting some space around one's thoughts to foster reflection.

Next, foregrounding materiality as a factor in the writing situation reduces the hypothetical nature of the conventional rhetorical situation and alleviates mindlessness in composing. That materiality includes the physical environment of writing—objects, texts, writing implements, qualities of temperature, light, and size—as well as the embodiment of the writer pertaining to any physical experience of writing—posture, gesture, breathing, bodily processes. Others have pointed to the unreality of accepted accounts of writing scenes, to the static Romantic depiction of the author's work space as one that "at once immortalizes and immobilizes the solitary writer" or for equating writing with solitude and the privilege of private offices, which are "detachable writing scenes" that cause disembodied, disconnected scholarly discourse (Brodkey 399; Fleckenstein, "Writing Bodies" 300). As a result of this placelesssness, "Scholars write as talking heads because they exist at the moment of writing as talking heads, displaced environmentally, corporeally, and rhetorically" (Fleckenstein 301). For first-year composition students, the scene of writing is more insidious than a garret mentality or a private study, because a hypothetical and mindless scene of writing occurs with every invocation of a rhetorical situation as conventionally taught. Students are taught to regularly project onto their mind a scene of a future time and place in which made-up characters of readers are reviewing students' writing, which is in a made-up, already polished condition. Trained to contemplate elsewhere, students learn to discount their experiences in the now, ones that could have made writing that ontological act of transformation Robert Yagelski seeks. In Bitzer's early construction, the rhetorical situation is a stage set for audience, exigence, and constraints, and it feels bare and abstract, with everything remaining a concept not part of the student's immediate physical context. A lack of awareness of the materiality of writing is the cause of the depersonalization of writing, as Kristie Fleckenstein and others point out; of poor thinking, as Ellen Langer ascribed errors in thinking to oblivion to actual context; and of writing apprehension, as Peter Elbow said happens to students who prematurely sign themselves up for meeting the expectations of challenging readers (*Writing with Power* 186–187). Conversely, restoring physical details and embodiment to the writing situation leads to holistic learning, stimulates invention, sets up realistic audience interaction, and allows the writing moment to become a source for connection with others.

Highlighting the material conditions of writing shows students how to better notice their audience projections, manipulate the proximity of their internalized audiences, and separate composing from editing. Post-Its, examination blue books, graphing paper, recycled paper, fancy stationery, and so forth are embedded with audience associations that can be manipulated to build a rhetorical context. A day-glow magic marker sets up a different rhetorical situation than a free bank pen or a Mont Blanc fountain pen. Keith Hjortshoj offers a memorable example of an undergraduate who struggles to complete papers in his first year at college due to an imaginary audience brought from home to his dorm room—his father, a professional writer, who had corrected the student's writing throughout high school. The student is finally able to turn in an assignment when he uses a sheet of crumpled paper; the marred paper lets him bypass his perfectionism, since his work was already "bad," and he altered subsequent assignment paper through scribbling or folding (Hjortshoj 15–17). This awareness of materiality while writing can prevent writers from slipping into hypothetical scenarios and direct students' attention to their actual relations and interconnectedness.

In Buddhist texts, the body is frequently a topic for the development of mindfulness, because it illustrates ways for relinquishing false attachments and for perceiving right interconnection. Body consciousness is separated into four Aggregates of Bodily Form: the Solid Element (including nails, teeth, bones, the liver, stomach, excrement); Fluid Element (blood, urine, tears, nasal mucus); Heating Element (food, beverage); and Vibrating Element (anything mobile or gaseous, including breathing and flatulence) (Goddard 24–25). In the Satipaṭṭhāna Sutra, mindful metacognition results from the contemplation of everyday physical movements: "a bhikku is one who acts in full awareness when flexing and extending his limbs ... when wearing his robes and carrying his outer robe and bowl ... when eating, drinking, consuming food, and tasting ... when defecating and urinating ... when walking, standing, sitting, falling asleep, waking up, talking, and keeping silent" (Ñāṇamolí and Bodhi 147). Practitioners are advised not to take delight in bodily forms or sensations but instead to recognize physical phenomena with a light, detached mindset (Goddard 27–29). Additionally, in the Second Noble Truth, the Buddha explained that clinging takes place via the body and our attraction to or repulsion from physical phenomena, that "[e]ye, ear, nose, tongue, body, and mind are delightful and pleasurable; there this craving arises and takes root" (Goddard 29). In Buddhist theory, the perception of physical phenomena leads to the dissolution of the autonomous self, replaced with a perception of interconnection. That is, material details show us our interdependence, regarded by Buddhists as dependent origination or "the notion that everything comes into existence in dependence on something else" (Lopez 29). No entity is self-created and independent, or uninflected by other entities, and to perceive the intrapersonal is to perceive that connected dynamic, the you in the I and the I in the you. Thich Nhat Hanh has called this interconnectivity "interbeing" or the "awareness that what we call 'I' is composed of 'non-I elements'" (Tworkov). In a similar fashion, Richard Shusterman says of this interconnectivity: "*A pure feeling of one's body alone is an abstraction. One cannot really feel oneself somatically without also feeling something of the external world*" (*Body* 70). As I sit at my writing desk, my awareness of arising physical phenomena might include the shiny black plastic of the laptop and my gratitude to the university for giving it to me, the slamming of a kitchen cupboard in the next room, the warmth of my turtle neck collar, all of which are items that connect me to others—whether my family or the factory workers and farmers who produced the items which I use. This focus on material awareness to establish a broader meta-awareness free of the trappings of ego can be found in contemporary contemplative pedagogic approaches that seek "connected knowing" and a view of the self "informed by its interrealtionalities and interdependencies" (Maitra 364; Wenger 24–25).

Material environmental details are advantageous for returning to the moment, but embodied ones specifically lead to a somaesthetics for composing. Somaesthetics, as defined by its pioneer Richard Shusterman, is the study of "how we experience and use the living body (or soma) as a site of sensory appreciation (aesthesis) and creative self-fashioning," ranging from analytic to performative and practical applications and including Zen meditation, exercise, and body sculpting, among other activities (*Body* 1; "Thinking" 14–16). With training, individuals progress through four levels of somatic consciousness, starting with basic consciousness without awareness and ending with the perception of consciousness itself—or mindful metacognition (*Body* 54–55). Christy Wenger's embodied imagination, enacted through her instruction of yoga in the first-year writing classroom, and Kristie Fleckenstein's somatic mind are examples of somaesthetics. These constructs surmount the dualism of mind/body that falsely isolates the ego, revealing interconnection and multiplicity, thus providing insight into the writing process and content for drafts. Both embodied imagination and somatic mind take an interface of body and mind as a way of interacting in the world—Wenger's embodied imagination a "creative fusion of the intelligent, organic body and mind toward the construction of present realities and future possibilities in writing" and Fleckenstein's somatic mind a "process continually making and remaking its own boundaries" (Fleckenstein *Embodied* 79–80; "Writing" 287; Wenger *Yoga* 21). Sondra Perl's work with felt sense is a much appreciated somaesthetic praxis in which students notice how insights for writing are often accompanied by visceral sensations, indicative of the role of the body in invention—widening the terrain of invention from just the head to the entire body. These approaches do much to diminish the tendency in composition theory and pedagogy to overlook the amount of physicality involved in writing. Students are diminished by that omission. They become apparitions in the moment of composing, not fully present, wavering in and out of view, disappearing into a future time or a past scene in which their writing is evaluated. It's this disconnection from the variables of the present-moment rhetorical situation that somaesthetics seeks to repair, replacing it with a view toward interconnection and dependent origination.

Conscious breathing has long served as an embodied meditation to heighten awareness of present experience. What better way to train the mind for awareness than picking an involuntary action, like breathing, something that would otherwise occur with or without our consciousness? The breath is a free, always available, highly portable learning tool for instruction and practice in the present. Breathing was one pedagogical tool in the Buddha's wheelhouse; he "taught specific antidotes to various faults" and "as an antidote to a wandering mind, he taught meditation on the breath" (Lopez 112). In addition to an entire sutra devoted to introspective breathing, the Ānāpānasati Sutra, a procedure for mindful breathing begins the Satipaṭṭhāna Sutra as practitioners are instructed to notice differences between particular inhalations and exhalations: "Breathing long, he understands: 'I breathe in long'; or breathing out long, he understands: 'I breathe out long.' Breathing in short, he understands: 'I breathe in short'" (Ñāṇamolí and Bodhi 146). The intent was to observe the moment but also to "tranquili[se] the bodily formation" (146). Christy Wenger does something similar with her composition students at the start of class; she helps her students "breathe their way into writing" by letting students select between three deep breathing techniques (varying the lengths of in- and exhalations) at the start of class based on how they're feeling that day (*Yoga Minds* 152). Through awareness of the rise and fall of the torso, passage of air through the nose, delicate patters of temperature change across the face and body, actions that are distinct to the moment, the practitioner develops a better connection to the kairotic present and is freed from the lure of the mind's pitter-patter. As Chögyam Trungpa writes, "each respiration is unique, it is an expression of *now.* Each breath is separate from the next and is fully seen and fully felt" (*Meditation in Action* 51).

In writing classes, attention to breathing, whether the breath-shaping practice of *pranayama* or the more observational practice of zazen, performs a powerful metacognitive function. The gains include changes in students' perception of their writing process, reflective classroom silence, and emotional flexibility (Wenger *Yoga*). The Zen practice of simply observing the breath without modifying it, as described by James Moffett, helps foster non-judgmental perception in the writing classroom ("Writing, Inner" 244). The impermanence of breath means that "it's always flowing; it's not a stable thing," according to Pema Chödrön, so that for the composition classroom, breath awareness is a natural pairing with low-stakes writing assignments (*How* 37). For Rudolf Steiner, disconnection from breathing leads to an overreliance on cognition, which in turn leads to disconnection from the world: "If cognition had been based on the rhythm of breathing instead of processes in the brain our whole relation to, and knowledge of, the world would be different" (41). Breathing has the potential to redress the dualism which Robert Yagelski says has plagued mainstream writing instruction, in which "writing enacts the Cartesian mind-body split by making the body irrelevant to the words the body produces" (45). Watching the impermanence of breathing in turn enables us to observe the impermanence of our thoughts, emotions, and perceptions. Following BKS Iyengar, Wenger believes that mindful breathing supports creative-rhetorical invention because focusing on the breath brings individuals to a dynamic situatedness. This dynamic flow of energy between self, objects, and other is ongoing and inventive, such that inhalation "literally opens us to new possibilities and ways of being and thinking that are in constant flux" (Wenger *Yoga* 159). By connecting breathing and writing, composition students practice a knowing and being in the classroom and at their desks that keeps them metacognitive, writing, and connected to themselves and others in the present moment.

A shift in consciousness happens when someone settles into the present moment to write. As the individual begins noticing the metronome of the breath, physiological changes occur, such as the slowing of the pulse and lowering of blood pressure. Watching her breath, the writer no longer chases after illusions about the nature of writing; she is less and less involved in a fox chase in the mind and eventually altogether a non-participant. In this fox chase, she ran ragged after a writing outcome, felt harassed by intimidating audiences in the head, and was goaded on by worrisome future outcomes. In addition to her breath, she begins noticing the real-time physical details of her writing environment, ones previously imperceptible although only inches away: warmth of the laptop, popcorn sound of the keyboard, creaking of the seat, resumption of the air conditioning unit, and the way the light hits the open composition notebook. This combination of physiological and perceptual changes runs in a loop, each recursively affecting the other (calm mind enables perception of real time; perception of real time calms the writer as she realizes her rhetorical situation is devoid of those stress stimulators). In Buddhist terms, she is less reactive to those stress stimulators and less likely to be commandeered by a storyline in which she would mindlessly respond to made-up scenarios, set in the future or the past, about the text at hand or her overall writing ability.

In Right Discipline, the mindful first-year student writer, dwelling not in the future, does not imagine a classmate holding his essay with a quiet smile on his face or with a puzzled expression. The mindful writer, dwelling not in the future, does not imagine a teacher moving a red pen about his work or inserting comment balloons. The mindful writer, dwelling not in the future, does not daydream about becoming a published writer, about reviews, about crowded seats at a reading, about thoughtful expressions, about moved expressions, about audience members approaching him afterwards with a question. The mindful writer does not see himself signing autographs. The mindful writer does not imagine telling his roommate about a great grade or a bad grade. The mindful writer does not envision herself opening an email attachment with a final grade from a professor and does not envision the "C." The mindful writer does not envision opening an email attachment from a professor and does not fantasize about the "A." The mindful writer does not visualize the pride of her parents or family at her writing success. The mindful writer does not see himself flipping through his computer files at the stack of accumulated work. The mindful writer does not imagine her essay at the top of a stack in her teacher's office; the mindful writer does not envision her essay at the bottom of a stack in her teacher's office. The mindful writer does not see a teacher setting down his book in disgust or stopping after the first paragraph. The mindful writer does not imagine the teacher showing his essay to the teacher's colleagues as an example of strong work from the class or as an example of weak work from the class. The mindful writer does not evoke the future, does not evoke the audience in the future, does not evoke her writing self in the future, does not evoke the writing of the future. Present moment, writing moment.

Interchapter 1

It's the First Day of the Semester

It's the first day of the semester, and students in my first-year composition course sit in as close to a meditative posture as one can in a computer lab, their hands resting on their knees, their eyes closed or softly focused a yard away. They're concentrating on the sensations of their inhalations and exhalations; the first assignment in the writing class seeks out the nonverbal. Every second or so, someone in the group—possibly me, since I participate in the activity—opens their eyes, moves their writing hand to a nearby pen, jots something brief on a scrap of paper, then lowers the pen and returns to the prior meditative posture. This continues for five minutes, intermittent bursts of pens clicking and clacking mixed with classroom silence of air conditioning unit and computer hum, hallway banter, coughing, shifting bodies. Afterwards, students share a cacophonic list with variations of "fut," "pas," and "eval" with some good-humored laughter at the predominance of one category in someone's list or just at the silliness of the group reading: *fut fut fut pas fut pas eval* or *eval eval eval eval eval eval pas*. My own list will tend to be skewed toward *fut*, since as the presumptive leader of the next class discussion, my thoughts turn toward planning what I am about to say. "When did mindfulness occur in the past ten minutes when you were doing the exercise? When were you aware of the present moment?" I ask. Students explain that making the decision to pick up the writing implement was the instance in which they were aware of the present moment—aware enough to see how their thoughts had deviated from watching the breath. "That was a moment then in which you were *mindful*."

The core training of a mindful first-year writing class is in present moment awareness: it must occur early (I start right after distribution of syllabi and introductions), and it must be sustained throughout the semester. The norm of mindlessness needs to be counteracted immediately and continuously to help student writers "remember to remember" and be aware of the present. In addition to amelioration of non-productive mindlessness, present-moment awareness is the fundamental lesson in a mindful composition course, because it is requisite for other components of process and rhetorical situation. From observation of present temporality evolve all the other lessons in the rhetorical present: intrapersonal rhetoric, affective dimensions of writing, flux, detachment, and emptiness. So in the classroom, we proceed from activities that guide attention to the present moment, to using writing to describe the present moment, and then to examining the present moment for the purposes of writing. Paralleling the Satipaṭṭhāna Sutra, the goal is to guide student writers from simple awareness to awareness of transience to detachment.

Exercises that guide attention to the present moment are typically short on written output, long on reflection, and always low-stakes informal tasks receiving little to no response or graded evaluation. At this early juncture, resultant class conversations do not necessarily link their encounters with mindfulness with the act of writing: it's sufficient to just practice noticing present temporality. These activities might include breath work, mindful walking, focused observation of an object such as a computer keyboard or a single letter on a keyboard, or list keeping of real-time changes in the classroom. The intent of this movement through exercises is to adjust students' focus to the present, to help them begin noticing how often they slip away from the present and the nature of those mindless occasions, and to experience first-hand the qualitative differences that come from sustained present awareness. We want students to collect first-hand experiences of present awareness and enjoy those experiences—see their personal and social benefits—as well as to begin to notice when their minds depart from the present moment. I make a solid case for the relevance of present awareness for them as learners, how it increases creativity, flexibility, and critical thinking by helping writers perceive multiple options and gain a less reactive, less judgmental stance on their own thinking (and writing)—sometimes by reading Ellen Langer. Again, they'll need reminders to observe the present, just as meditators might use a gong or a common event like touching a door knob to return them to the present. Finally, these initial exercises can't be one-off but must instead be repeated throughout the course to maintain mindfulness. For instance, the following "Mind List" activity can be used as a prewriting device whenever they're given a fresh assignment as well as throughout other phases of working on a piece of writing, including revision. In this case, students pair the breath-watching activity with a focus on a draft or even on a particular formal or content area. With inhalation and exhalation, they direct their minds to that area of their writing to obtain new insights and gain a sense of their evaluations and preconceptions: lists of past-, future- or evaluation-driven thoughts collected would concern how students are perceiving the writing task at hand.

The praxis I find effective in helping students gain a first-hand experience of mindfulness is that list keeping activity, "The Mind List": it tracks their departures from nonconceptual attention to their breathing. I like this exercise because it gives most students that first-hand experience of mindfulness, while other activities might leave some students frustrated, believing the concept of the present too abstract and feeling they've been left out of the "enlightenment" or "ah ha moments" their peers seem to be enjoying. As mentioned in the opening to this chapter, students place a piece of paper and a writing implement close by. I ask students to sit gently upright, hands resting either palms up or palms down on their knees. I ask them to scan their body for tensions and sensations. With eyes gently focused on a spot a few feet away, students begin to watch their breathing, noticing the sensation of breath as it enters and exits the nose, noticing the rise and fall of the torso. (I often use the basic mantra, "Breathing in, think to yourself, 'Here.' Breathing out, think to yourself, 'Now.'") I ask students to continue to balance on that attention to breathing, but whenever they notice that their mind has wandered from watching the breath, they should turn to the piece of paper and jot one of the following: FUT (for a distracting thought about a future moment after this meditation); PAS (for a distracting thought about a time before this meditation); or EVAL (for any thought that judges the present circumstance—for instance, whether the student is pleased, irritated, or bored with the now). This notation is done fairly quickly, without making a fuss about it. They're not to elaborate in their write-up (i.e., no need to describe the nature of the future-oriented daydream or the judgment) but to simply and without self-recrimination replace the pen to the notebook surface and return to watching their breathing until the next monkey thought arises. The next time students discover that they're no longer watching the breath, they return to the paper and again record which category they've caught themselves using.

After five or ten minutes, each student's paper is probably a long laundry list of distractions. The clattering and clicking of pens is prominent—the teacher's as well. I tell students that wandering off the breath is inevitable and that they should hear many of their neighbors' pens being engaged. After three or four minutes, I ask for a few volunteers to read aloud their list without providing details: it often makes for a humorous syncopation of "fut" "fut" "pas" "pas" "eval." Students are usually amused by the inner discursivity or monkey mind occurring all around them in the consciousness of their peers. After highlighting the free-ranging quality of inner talk, I point out that they've had an encounter with mindfulness, since the split second in which they realized they'd been daydreaming or evaluating and moved to pick up the writing implement was a moment in which they'd been aware of the present moment and how their discursivity occupied that moment.

The key is how students experienced mindfulness the moment they realized they were engaged in future, past, or evaluative thinking and picked up the pen. What lead them astray from that attention to breath was their inner discursivity. Mindful metacognition happens in this activity when students first notice that their attention has wandered off the breath, and it continues as students critically examine the nature of what has lured them off the breath. In the last part of the activity, students return to watching the breath. As a group, we discuss the qualities of that mindful consciousness: how it's a different type of awareness, one specific to the moment at hand. We discuss our human proclivity for sorting and evaluating experience and how that near-non-stop judging might impinge upon our writing. A side effect of this activity is an increased awareness of one's surroundings—the drone from the motor of the overhead projector, hallway sounds, rustling of clothing, swallowing.

Exercises that use writing to describe the present moment work with the senses and evocative detail to practice the contemplation of "body as body," "feeling as feeling," and "thought as thought" mentioned in the Satipaṭṭhāna Sutra. The point is to expand students' awareness of the present through sustained description, making the present moment the object of study by pairing awareness with language. Mindful metacognition transpires with contemplation of everyday physical movements and gestures in these frequently embodied writing activities—the physical side of writing plus mundane steps such as turning on one's laptop, done with mindful awareness. The goal is to help them become less reactive and more detached: to gain some critical distance. These exercises tend to be more formed than ones simply striving to provide students with experiences of mindfulness, and they will involve more writing. For instance, students complete "10 Breaths" in class by observing their breathing and describing each breath as an objective experience—its size; volume; interactions with ribs, collar bones, and other parts of the body; temperature—as well as figuratively through metaphor synesthesia (when different colors or textures are associated with a particular breath). I offer prompts as students observe their breathing: What color is the next breath? If it were an object, what kind of object? One breath may bump against the ribs; another breath looks like an enormous green bottle; another breath has an ambiguous object lying inside of it. Probably any type of embodied writing will work well: describing the sensations of posture, clothing, or furniture while they write. The activity that I frequently assign is one of mindful eating.

With "Mindful Eating," I provide two snack items (usually a combination of natural and manufactured food such as a tangerine or apple slice and an Oreo or Triscuit) in a variant of Jon Kabat-Zinn's widely used "eat one raisin" exercise. I walk students through an initial mindful breathing activity, and I instruct students to continue to watch their inhalation and exhalation as they pick up one snack item and examine it with their eyes, noticing colors, light reflections, shadows, structural formations. Students may be asked to record these observations, share them aloud, or simply think them. Next, students rub a finger across the surfaces of the item while observing the breath. Next, students smell the item, simultaneously observing their breathing. Then I ask them to imagine all the places the components of the snack (wheat, sugar, salt, etc.) had traveled prior to the classroom and the different people possibly involved in its growth, harvest, manufacture, and shipping—while watching the breath. Finally, students take one bite of the item and observe the sensations in concordance with the breath, chewing at least twenty-four times before swallowing. Students report an almost surreal perception of the food and the experience of chewing and swallowing—an activity they've done on autopilot in all likelihood several times already that day. The Oreo becomes intricate as students notice an elaborate design on either side of the cookie, one that includes both writing and drawing. The Triscuit suddenly possesses a sandy surface and an unpleasant razor's edge. Returning to this activity as homework, Phoebe selected an almond and "noticed that it had a strong nutty smell that reminded me of nature, of being outdoors gardening." A piece of breakfast cereal called "Froot Loops" felt like a "pumice stone" and when it was raised to the nose, Jenny "could not pick up a specific fruit that it smelled like." For Tatiana, a popcorn kernel in her dorm room "resembles a tiny cloud with a tiny yellow center that seems to get darker and darker" and when she applies pressure to the popcorn, she feels "the single piece of popcorn multiply into five pieces." The act of eating even becomes monstrous—ocean sprays of saliva, bumps against the back wall of the mouth, overwhelming tastes. Some students have actually become repulsed by the experience of mindfully eating a prepared food item, their mouths, they note, awash with chemicals and artificial flavors never previously detected. Attention is drawn to an object and activity in the moment, and it becomes startlingly vivid, unique, and evocative, with a plethora of detail. To defamiliarize a familiar and usual rote act carries learning lessons in awareness that can spill over onto other activities such as writing.

This combination of observing and describing the present moment resurfaces in the mindful composition course as an invention method or way of engaging in prewriting with "yoga for hands." The heuristic advances student writers' knowledge of verbal emptiness, a concept I'll discuss later in Chapter 4, double-dutying in its use of students' embodied knowing (inherently less verbal to fully nonverbal) to develop the calm, perceptive outlook needed to make the leap to producing new wording. "Yoga for Hands" heightens writers' present rhetorical awareness by asking them to pay attention to their fingers, hand, arm, and other parts of the body while they handwrite or type—the often overlooked flurry of activity and sensation through which writing occurs. More often than not, I maintain, thinking about our fingers as we type would benefit us more than thinking about a hypothetical, future-situated reader. The heuristic, like any of mindful invention, establishes a present-focused state in students that is free of audience taxation. It's really wondrous how much activity—the complexity of the physical movements of writing with a hand, the various sound effects like typing or the glide of a wrist along a sheet of paper, the smell of ink, the warmth of an overheating laptop—surrounds even the ordinary act of typing, often without our conscious awareness of these sensations as part of our writing environment. In Phoebe's account of yoga for hands, it was "like a million bees were rushing through my head. While I was writing, the keyboard kept making a clicking sound a mile per minute, and I was typing non-stop as though my hands were on fire. My fingers were on top of their game, my hands and fingers were in perfect harmony, my arms were resting against the keyboard, and my shoulders were relaxed." By the end, she states that "I came to realize that I'm a relaxed writer, and I became in tune with how my body and thoughts are connected while writing." Marcus remarks on affective developments from the method, saying, "It feels like I am accomplishing something important, and I move my fingers and click the keys. The clicking feels good on my fingers for some reason that I cannot explain." Jonah reaches this comparison: "When my fingers touch the keys, it feels like I am touching a floor, a floor that I can push. It's like each different floor I push types a different letter on this document." Before doing yoga for hands, Jonah says, "I have never thought of how I physically felt when my body types a document." Students make the subject of their freewriting their own bodies, and they circulate between the embodied knowledge of felt sense and the generation of words about the body. In turn, that generated text promotes bodily consciousness, which results in more text, and the happy cycle continues. By the end of the heuristic, students can be encouraged to consider upcoming writing tasks.

The method begins with a brief seated meditation in which students draw attention to the breath. Next, students move their hands to a keyboard or piece of paper and commence with an unfocused freewrite while maintaining breath awareness. In mid-stream of this freewrite, students turn their attention to the sensation of their fingers touching the keys, pen, or pencil and change the topic of the freewrite to describing this sensation for a minute. The focus of the freewrite changes to the sound effects of typing or handwriting. Next, students sequentially move their attention from the bones of their writing fingers (watching the complexity of their activity) to the palm, back of the hands, wrist, lower arms, torso, legs, shoulders, neck, and finally the face. Concerning the face, students notice how the act of writing affects its muscular movements, tensions, and changes in temperature. Afterwards, students are guided to continue the observation of the breathing and to return to the day's project.

The next of the three outcomes sought in mindful metacognition, namely, awareness of transience, takes the form of low- and higher-stakes activities throughout the semester. While other teachings related to the Satipaṭṭhāna Sutra help student writers think less frequently about the future or past as concerns their writing, the intent of the next outcome is to help students notice and appreciate how no two writing moments are identical. The benefits of this constant change are multifold, including the reliable provision of possible content for pieces (all those shifts in intrapersonal dialog lead to an abundance of material if observed with detachment) and a sense that the writing self and its abilities are not static and predetermined but instead dynamic and brimming with possibilities. The constant change in the rhetorical writing moment is a student writer's resource, not burden. Impermanence affects every other rhetorical factor, including self-ethos, exigence, and assorted attitudinal constraints—as will be discussed in a later interchapter. Freewriting, of course, is a marvelous method for tracking changes in the mind and its intrapersonal rhetoric, and student writers can be asked to complete two private freewrites during the same class meeting, either open topic (what's-on-your-mind-right now) or focused on a particular writing project, and then examine each to locate subtle shifts in outlook, idea, and physical sensation.

Along with facilitating content development, the perception of impermanence suits revision work if revision is stipulatively defined as "to continue to perceive change," with less evaluation of that change on the basis of its quality. The moment revises itself, so there is no need for a forced attempt to set out and revise, no need to make revision a discrete phase from the rest of mindful composing. The changing moment, if perceived, is always handing us revisions. Because impermanence can seem beyond our control, introducing as it does a host of alternatives, angles, alterations, it can appear to run aground of the stability we seek in making a piece of writing. Impermanence can be challenging to embrace if students fear evaluation and grading, because tracking impermanence means little to no time for fixing, correcting, revising, or polishing. Consequentially, assignments need to accommodate impermanence and value the perception of flux as a learning outcome, such that even high-stakes tasks reward open-ended products. A piece of writing may continue to change well past its eventual final, polished form; the best version may occur somewhere in the middle or near the end of a writing process. Different drafts are described without evaluative language—this is what this draft does in form and content—to circumvent our tendency to sort change in order to make ourselves feel more comfortable. Assignments focus on increasing students' facility with impermanence through two tracks: the first dwells on the ability to observe and utilize constant change to affect a draft; the second concerns the ability to produce quantity rather than quality (quantity as the tracking of that constant change) to build a draft. This second focus can tremendously benefit first-year writing students, since it builds their faith in an abundance of writing material, and when writers trust that more material is just around the mental corner, they will be less resistant to changing existent writing, less clinging, and more mindful. In general, assignments working with rhetorical impermanence strive to develop students' acceptance of change and their stance of detachment.

I assign excerpts from Raymond Quineau's classic *Exercises in Style* and ask them to experiment with radical revision on a passage using one of his ninety-nine methods (i.e. doubling words, adding conversational filler, telling a narrative backwards in time). In another prompt, students brainstorm for ten different openings to a piece: an image, a list, a question, on dialog or code switching, a comparison, an example, narrative, a fragment, an address to the reader, an exclamation. I have found it useful to teach a variation of a Don Murray exercise on late-stage drafts or even final drafts in which students brainstorm for twenty variations on the piece: large-scale variations in genre and audience while maintaining the same content. As an example, a scholarly article could morph into an article for a trade journal could morph into an interview could morph into a poem could morph into a blog posting could morph into a personal essay could morph into a Tweet could morph into a series of process notes. The message sent is that even polished pieces undergo change and that invention can occur long after the last spellcheck. Similarly, students come to class with a sentence or paragraph from someplace in a draft isolated on a sheet of paper, and other students in the class take a guess at what might precede and follow the excerpt in terms of content and structural choices they'd like as readers.

To reconfigure the constructs of process and situation to increase mindful metacognition, we examine common conceptual metaphors for composing. We discuss how conceptual metaphors structure our everyday functioning by organizing our perceptions and actions—how they're "pervasive in everyday life, not just in language but in thought and action" (Lakoff and Johnson). Conceptual metaphors are large conglomerates of thought, or as Philip Eubanks describes them, "metaphor expressions [that] recruit larger metaphoric concepts"—such as "Time is Money" (44). Behind this seemingly benign three-word metaphoric phrase is a belief that has the potential to shape how you use your next hour or even the rest of your life. In composition pedagogies, the terms regularly used to talk about writing—freewriting, brainstorming, drafts, revision, feedback, prewriting, outline, proofreading—transport unrecognized assumptions about what it means to write. If we take the connotation of "draft," for instance, we might list breezy, temporary, fleeting, insubstantial, invisible (notice only its effects), although a graduate student also said draft as in a draft beer (a volume, abundance, a small sample from a much larger supply). I ask students, *What are the connotations of those ordinary words about writing? What sorts of imagery do they contain for you? How might those ordinary words impact your writing experience or writing outcome?*

To determine the impact of those connotations, students generate synonyms for a conceptual metaphor. For example, a synonym for "draft" is "stage." How is "stage" different in what it suggests from "draft"? By finding alternatives, we can better notice the original language. If we carry around the idea of an early draft as fleeting, this could correspond nicely with the sense of impermanence, if we are of the Buddhist mindset. That breeziness or invisibility, however, could make early writing seem hard to catch and increase the difficulty of starting out on a piece. Then again, if draft is like pulling from a keg, this suggests an abundant inner supply. An alternative to this assignment is to ask students to create new conceptual metaphors for various parts of the writing process and rhetorical situation. After crafting a new term, students might do a freewrite or momentwrite to unpack their newly minted conceptual metaphor, exploring its possible usage in writing practice.

Another activity asks students to create one-sentence mantra for each course concept (for present awareness, for intrap, for emptiness/prewriting, for affect) or as a way for students to summarize learning in a way different from the usual metaphor of process of situation. My own mantra is "Your ability to write is always present." Mantras created by my students include "The writer is always the block"; "In understanding how to proceed, there must be delay"; "In the teaching of groundlessness, this sentence will be read differently by every other person: it will never be the same sentence"; and "Realize that there's no ground for the writing block to sit on, and you will soon realize it will not be blocking for much longer." Mantras from my most recent first-year composition students include: "Your writing is already enough the way it is" (Amanda); "Focus on the now" (Claudia); "Prewriting can solve all writing headaches" (Taylor); "If you can't write, just breathe and try again" (Jenny); "Freewriting is always a good prewriting" (Jonah); and "Stay in the only moment" (Marcus).

2

THE MONKEY MIND OF INTRAPERSONAL RHETORIC

> Something
> Ought to be written about how this affects
> You when you write poetry:
> The extreme austerity of an almost empty mind
> Colliding with the lush, Rousseau-like foliage of its desire to communicate
> Something between breaths, if only for the sake
> Of others and their desire to understand you and desert you
> For other centers of communication, so that understanding
> May begin, and in doing so be undone.
>
> —John Ashbery, "And *Ut Pictura Poesis* Is Her Name"

Every time a student sits down to write, two texts happen nearly simultaneously. One text comes with a font and a page appearance; it's the one that's revised and distributed, uses sentences and paragraphs. It can be spellchecked and reread. The second text is not disclosed and operates invisibly, often outside even the writer's notice. This one is continuously moving on, never revised, an array of fragments that wildly diverge in content: this second text does not adopt the form of a draft. Nevertheless, the wording of this text exerts powerful rhetorical influence through pathos, connotation, imagery, and sometimes logos, affecting content and process moves, and the recipient of its dubious attention is the student writer. Eventually, some of this internal wording might manifest externally in the visible text, but most of its content—its asides, meta-commentaries, registering of physical sensations—disappears with the next moment. This invisible text is constantly being produced; in fact, it's difficult to shut off its production because of the geyser of discourse in our minds. The first text is part of an external, interpersonal rhetoric that has been taught and analyzed since the dawn of rhetorical education; the second text is part of an internal, intrapersonal rhetoric that always occurs alongside the external and is, in fact, formative to it.

This chapter examines the near-constant arising of inner discursivity that occurs in human consciousness as a factor in a rhetorical situation—intrapersonal rhetoric. Intrapersonal rhetoric is the most immediate discourse available to students: first on the rhetorical scene, its shaping influence on subsequent external rhetoric should not be underestimated. All writing, no matter what the genre or audience, *all writing* begins as intrapersonal communication, despite how we continue to dwell on the interpersonal future, looking for the essays, pages, grades, publications of upcoming moments—overlooking the language produced immediately in front of us. The intrapersonal is the material that consistently fills the majority of newly arising writing moments. Since inner discursivity occasionally leads to finished texts, and since finished texts are always dependent on that inner discursivity, we would be wise to ask what might be gained by better observing intrapersonal rhetoric. The chaotic influence of intrapersonal communication is picturesquely captured in the Buddhist notion of "monkey mind" or our non-stop tendency to sort and evaluate experiences in our heads—yet another reason for working on a mindful metacognition of self-talk inside the writing moment. Not entirely a writer's benefit, the intrapersonal foments problems by generating preconceptions and other writing liabilities, in addition to purveying content. As a result, the intrapersonal is both a low- and a high-stakes event. It's low stakes in the sense of a frequent, unrevised, and private communication; on the other hand, the intrapersonal is high stakes in that it's hugely influential on the writing situation, impacting writing self-efficacy and perceptions of audience. In this chapter, I discuss writing as a private, internal discursivity that takes on the paradoxes of self and other, ones illuminated by Buddhist theories of no-self. Learning to control inner rhetoric through mindful metacognition benefits student writers in a myriad of ways—not the least of which is its confirmation of writing ability and dissolution of writing blocks.

While claiming the originary nature of this internal discourse, I am not implying its originality: intrapersonal rhetoric is markedly intertextual, carrying the fingerprints of other writers and thinkers and our verbal inheritances. From a Buddhist mindfulness perspective, to claim otherwise is to perpetuate a false sense of self and other as discrete entities. Intrapersonal rhetoric is comprised of intertextual, socially inflected instances of language that nevertheless occur in solitude as the student writes in a singular present moment. Intrapersonal rhetoric is intertextual because it pirates non-stop from prior occasions of literacy—those "citations, quotations, allusions, borrowings, adaptations, appropriations, parody, pastiche, imitation, and the like" that connect one text to another text, one writer to another writer (D'Angelo 33). Kristie Fleckenstein describes intrapersonal rhetoric as that "discontinuous stream of internal monologues" which transports culture to the mind through discursive codes ("Writing Bodies" 289–290). Many phrases arrive between invisible quote marks—recalled conversations, song lyrics, an author's turn of phrase. In contrast, from a mindfulness perspective, interpersonal rhetoric occurs *only* when a reader and a writer's text dwell in the same present moment or, in the case of spoken communication, one person directly speaks to another. The intention to share a document or the preparation for that sharing do not constitute moments of interpersonal rhetoric. For instance, although I hope that the previous sentence clearly communicates to someone at some point, when I typed it ten seconds ago, I was alone. The actual real-time activities consisted of a person (me) typing, listening to Dr. Jeffrey Thompson's *Music for Brainwave Massage* through Bose headphones, at 3:04 pm on February 15, 2017. Social media genres like texting or emailing make good demonstrations of this distinction, since in their immediacy of delivery (and potential for response nearly as quick as a face-to-face conversation) they bolster an illusion that they are interpersonal. However, if we save a draft of our email or delay sending a text for a few minutes, we can put the brakes on the illusion that we have been speaking to someone as we typed. Therefore, the *material* of the intrapersonal—its bits of voice, phrases, sentences—is intertextual and informed by the language of others. At the point of utterance, the actual *address* of the intrapersonal is directed to no one but the self; that is, the *actual* or *present* audience of the rhetorical situation is the self.

It may also seem that what I am proposing about the primacy of the intrapersonal represents a regression in the discipline of Composition Studies—a suggestion that we equate writing with thinking. It might seem that I am coming hazardously close to suggesting that students refrain from writing, from producing anything that could be shown to readers, until they have mastered their thinking. Such a notion would undoubtedly constitute a reversal of the achievement of the process moment in the 1960s and 1970s in its counteraction of get-your-thinking-right-before-you-write, its permission of student error and exploration through drafting. Pre-process textbooks and instruction often mandated that students straighten out their thinking: a "think, outline, write, revise model" that advised a defensive planning to anticipate problems, a model which continues in the requirement that students develop a thesis before drafting (Newkirk *The School Essay*; Tobin 3). While I am equating writing with thinking in that it requires undisclosed, invisible discursive activity, I am not issuing judgments about the quality or relative worth of students' thinking and, relatedly, am not at all proposing that students keep an audience in mind as they work intrapersonally.

Instead, I suggest that the several benefits that result from viewing the intrapersonal, or undisclosed discursivity, as the basis of writing include the acceptance of impermanence, the recognition that the intrapersonal carries real internal and external consequences, and the understanding that writing doesn't require anything extraordinary beyond the ability to remain as non-evaluative as possible. Identifying internal discursivity as a verbal activity equivalent to more valorized externalized writing alleviates writing apprehension and signals that this verbal activity is worthy of metacognitive exploration. When the intrapersonal is taken up as a type of writing, it means that writing can be so fleeting as to remain unrevised and neither audible or visible to others. Like disposable writing, unrecorded intrapersonal communication will not be read by others or rescanned by its author, who may retain insights, stances, and even phrases from it.

Finally, in making my case for the primacy of intrapersonal rhetoric, I do not overrule other present rhetorical factors such as impermanence or materiality, both of which are in interplay with inner talk. From a Buddhist mindfulness perspective, it is important to avoid perceptions of phenomena as discrete and autonomous, for to do so is to strip phenomena of time and connectedness. As will be shortly discussed, "[a]nything that exists autonomously, independently, or objectively can be said to have a 'self'" that is illusionary (Lopez 29). Intrapersonal rhetoric arises and unfolds in a context, one of present-moment time; its continuation depends upon those temporal fluctuations. Similarly, while it certainly may engender preconceptions, misperceptions, and emotional responses during the act of writing, the intrapersonal is also subject to influence from those rhetorical factors. For instance, a first-year composition student's thoughts while writing may veer toward sharp worry about an intimidating instructor, and that worry will then dictate the next set of words that emerge from his or her self-talk. That's why part of mindful composition pedagogy entails helping students to perceive intrapersonal rhetoric in relation to other factors. A writer who fixates on internal talk will not be able to optimally make use of its insights and material, which unfold through impermanence and occur through material realities, if it's stripped of present time. However, of the various rhetorical factors in a mindful writing situation, intrapersonal rhetoric most resembles the final product writers seek and can contain similar words, phrasings, and sentences. And yet, affective and physical dimensions of the writing experience will not directly make it into the final copy. The taste of decaf espresso on my tongue and front teeth right now would not normally be shared with a reader as content, unless evoked through description, as I just did.

Characteristics of Intrapersonal Rhetoric

As the language of the moment, intrapersonal rhetoric is both highly persuasive and ongoing. The mind doesn't suffer from roaming charges because intrapersonal rhetoric is ubiquitous and nearly continuous, a combination with considerable implications for writing ability and task environment. This rhetoric is intertextual on the level of the phrase and word, comprised of a heteroglossia of recalled social instances of language from any genre, using words coined by others (otherwise every second of writing would require the creation of neologism). Bakhtin put it this way: "The word in language is half someone else's" (293). Intrapersonal rhetoric can manifest in two forms: it can be simply observed as part of thinking or written down and steered through metacognition, made epigenetic, as is the case with freewriting. For the purposes of mindful composition, I generally prefer "intrapersonal rhetoric" over other terms, including Jean Nienkamp's "internal rhetoric," because intrapersonal rhetoric suggests a dialog with the self, an internal

Q & A or call & response, that requires critical examination of one's unvocalized thought. As established by Vygotsky and Piaget, the intrapersonal is integral to human development, such that study of the intrapersonal comes with a long-term, developmental view rather than anything expedient, useful for an assignment or course. The intrapersonal calls for the physical—not metaphoric—silence of the individual and occurs in a separate time and place from interpersonal rhetoric and potential future readers. Finally, the intrapersonal is common. It's not a special ability: everyone generates it (we can't help it), although remaining aware of intrapersonal rhetoric is another matter and requires training, practice, and discipline. That the intrapersonal is ordinary is a point that cannot be overemphasized, because the intrapersonal, like breathing, is a shared capacity and a regularly available resource for writers. Readers of this paragraph can immediately perceive this fount of verbal production in themselves by engaging in any sort of nonverbal or erasing activity, one that shifts attention to a physical sensation like breathing. After a few seconds of nonverbal consciousness, bits of discursivity start to surface on the mind, almost as though attention to the nonverbal has summoned the verbal. As I was flicking filaments of eraser just now from the tip of my mechanical pencil, focusing on my breath and on my fingers and the pencil, "speech," "you can often," "for instance," and, inexplicably, "Angelina Jolie" scribbled on my set-up moment of blankness.

In *Internal Rhetorics: Toward a History and Theory of Self-Persuasion*, Jean Nienkamp argues for greater scholarly attention to the intrapersonal as a little-recognized influence on our thoughts and actions, with powerful implications for writers. Nienkamp calls for the study of internal rhetoric, since "talking to oneself is at least as ubiquitous and consequential as the various social language practices analyzed from a rhetorical perspective" (ix). She points to the highly persuasive nature of self-talk—how it influences moods and actions on a regular basis and flies under the radar of most people's consciousness. In Nienkamp's historical review, Plato's separation of speech and thought initiated the dismissal of internal rhetoric, a move reinforced by Aristotle and then complicated in the twentieth century by the notion of the unconscious (exemplified in the work of Kenneth Burke, Chaim Perelman, Lucie Olbrechts-Tyteca) and the trajectory of human development through language use (exemplified in the work of George Herbert Mead and Lev Vygotsky). According to Nienkamp, individuals possess rhetorical selves, which are the "ongoing product of individual acts of internal rhetoric," and because thought so deeply shapes our minds, it is rhetorical (x). She suggests that the machinations of internal rhetoric dodge our everyday and scholarly attention because of yet another trick of the mind, the "internally suasive power of our accepted categories," and echoes Ellen Langer's diagnosis that mindlessness happens with reliance on previously established distinctions (Langer *Mindfulness* 11; Nienkamp x). A particular type of internal rhetoric, "primary internal rhetoric," results when the unconscious persuades us, typically through an internalization of societal language (Nienkamp 113). In her view, the usual binary of interior/exterior or individual/social and standard rigid categorization of some composition scholarship is not a factor, since "internal rhetoric involves interiorized social voices similar to those that shape external rhetoric" (127). As a result, the rhetorical self is "not a unitary self" but, rather, a repertoire of voices, a point which will shortly become important to understanding the Buddhist sense of no-self in mindful composing. It's through self-talk that we cultivate our internal rhetoric and sculpt our rhetorical selves: or a cultivated internal rhetoric that we may become—a useful model for proceeding with a Buddhist mindfulness practice of the intrapersonal.

Working with the intrapersonal doesn't always involve writing things down, and yet it remains a controlled inner rhetoric. Not all intrapersonal rhetoric becomes recorded on the page or screen: some of it is simply overheard by the writer, though it is nevertheless influential on self-efficacy and content. Emily Dickinson said, "The Spirit is the Conscious Ear." In describing an internal reading down by writers as they write, or a "sophisticated reading that monitors writing before it is made, as it is made, and after it is made," Donald Murray comes close to the intrapersonal ("The Essential" 88). This internal reading entails the perception of emerging language by the self, presumably some of which is unrecorded and pure thought, since it is seen "before it is made" and the writer reads "what isn't on the page as well as what is on the page" (88). Observed internal language is low-stakes practice in relinquishing verbal material, preventing attachment to outcome and product. Strictly observational intrapersonal rhetoric happens through a student's intent to just sit and watch inner discursivity (not type anything), but it even happens if the student's intention is to create a record of passing thought—not every nuance will be caught. Because a portion of intrapersonal material invariably eludes a written record, there's a quality of private writing to it, no matter that it borrows from social language practices, because "[e]ven if 'private writing' is 'deep down' social, the fact remains that as we engage in it, we don't have to worry about whether it works on outside readers or makes sense" (Elbow "Closing" 105). Just as a bag of flour is not synonymous with a cake, intrapersonal rhetoric is not the same as interpersonal rhetoric—or, indeed, not the same as a draft or even a freewritten transcript. Part of the bagged flour (paper container, dead ant in its white dunes, substances adhering from kitchen counter to bag bottom) does not make it into the cake, and not all the flour will be used in a standard-sized dessert. Similarly, a piece of writing is never a direct transcript of intrapersonal experience, which is often a complex admixture of inherited language, commentary on physical sensations that happened while writing synchronously with breathing, or reactions to stretches of wordlessness.

Intrapersonal rhetoric bears comparison to freewriting and to stream of consciousness, two concepts also concerned with noting quickly changing thoughts. Freewriting is a written record that tries to yoke passing thought to words on a page or screen. It tries to pick up the pace by giving the writer his or her marching orders—a person doesn't wait for inspiration or ideal circumstances to freewrite and is told to "simply force yourself to write without stopping for ten minutes" and to "accept this single, simple, mechanical goal of simply not stopping" (Elbow, *Writing with Power* 13). Freewriting seems almost athletic because it joins the physical activity of handwriting or typing with thinking, leading to a faster pace. I think this is why I frequently use the Nike slogan, "Just Do It," when talking about freewriting with students. Keith Hjortshoj says freewriting heightens ease through its coordination of psycho-physical activity: "[I]magine that the mental and physical aspects of writing are two wires. By removing all of the concerns about the product that create second thoughts and hesitations … [t]hinking and writing become a single, uninterrupted activity, both mental and physical" (29). The physicality of the activity and the tangibility of the results of a freewrite exert influence back on freewriting: since freewriting is visible, it carries more resemblance to interpersonal writing, because its physicality and physical presence invite the chance that it *could* be read by others. Through mindful composing, students scan the intrapersonal for arising rhetorical factors that wouldn't make it into a draft (i.e. one's preconceptions about the writing task or physical experiences while writing). At an advanced stage of mindful writing practice, a student strategically draws attention to those passing, unrecorded bits of intrapersonal to return his awareness to the present, much as he might have selected his wrist movements or breathing for the same purpose. Furthermore, freewriting requires that the writer perform two simultaneous activities, namely, observing internal language and moving the body to record that language, unlike observed but unrecorded intrapersonal and stream of consciousness, where there's a single activity (watching).

Like Elbow's freewriting, William James' notion of a stream of consciousness underscores quick change, though with more emphasis on impermanence and interiority. Thinking is in effect defined as internally experienced flux, because "within each personal consciousness, thought is always changing" and is "sensibly continuous" (225). In making his case for a stream of consciousness, James raised points similar to the ones Ellen Langer made for mindfulness. Change in mental experience is a constant, and it's a mistake to search for uniformity in either our sensations or our thoughts, since "our state of mind is never precisely the same" and a "permanently existing 'idea' … is as mythological an entity as the Jack of Spades" (233; 236). James' rejection of freestanding, everlasting, and therefore static ideas is in accord with Buddhist notion of dependent origination; the mutability of the mind seems a perfect set-up for the creation of new categories and search for novelty Langer ascribes to mindful learning. James' conceptual metaphor of a stream of consciousness picks up on Heraclitus' notion of dissimilar experience, with the caveat that it's a perceived rather than actual continuous flow:

> Consciousness, then, does not appear to itself chopped up in bits. Such words as "chain" or "train" do not describe it fitly … It is nothing jointed; it flows. A "river" or a "stream" are the metaphors by which it is most naturally

> described. *In talking of it hereafter, let us call it the stream of thought, of consciousness, or of subjective life.*
>
> *(239)*

However, consciousness only appears continuous to us since it's frequented by "time-gaps" (both unfelt and felt) and "breaks in quality": the former referring to how individuals exit consciousness for stretches (sleep as the obvious example) and the latter to swift changes in topic in thought (237). The thought procession is infiltrated by other non-discursive moments, including "transitive parts," or the cognitive moments which escort us, through considerations of relations, from one area of contemplation (or "substantive part") to the next (243). The comparison James uses for transitional moments is a snowflake: elusive, since as soon as you capture it in your hand, the moment has changed (248). In addition to this assortment of breaks in consciousness, there's also the gap of namelessness, evaluation, and "psychic transitions," which differ from transitive parts in that they are more premonition-like—awareness of thoughts that have yet to be articulated (251; 253). James diffuses the potential for a hierarchy of consciousness between the more three-dimensional ideas and the blanks. In a manner that harkens to the Buddhist interplay of form and formlessness, he says that "namelessness is compatible with existence. There are innumerable consciousness of emptiness, no one of which taken in itself has a name, but all different from each other" (252). While freewriting is dedicated to the verbal, stream of consciousness acknowledges a mix of ideas and gaps and in this regard is similar to intrapersonal rhetoric, which is a mix of affective and physical rhetorical factors as well as words.

The ability to notice intrapersonal rhetoric is linked to improved writing and learning experiences, and, conversely, not perceiving intrapersonal rhetoric carries negative consequences for social interactions (including learning). Others have described how intra- and interpersonal are mutually dependent, a dynamic crucial for meaning-making with others and for personal intelligence (Barbezat and Bush 15; Gardner 291–292). Intrapersonal rhetoric, however, as the actual rhetoric of a present rhetorical moment is the primary means to achieve better writing experiences and outcomes. Overlooking the intrapersonal causes breakdowns in interpersonal communication, as Carl Rogers, founder of client-centered psychology, proposed in the psychoanalytic context, and by extension, writing difficulties are attributable to the failure to notice inner speech. In "Communication: Its Blocking and Its Facilitation," Rogers argued that a patient's miscommunication with others was symptomatic of hidden ruptures in the patient's communication with self and that a therapist could help the client restore self-communication by listening without evaluation. Students seldom have been encouraged to operate in this mental "space" in which their writing is accountable to no one. From students' earliest educational moments, their literacy has been largely supervised by internal monitors established by reminders that they keep in mind audience needs and expectations, reinforced by looming grades and evaluation. As a result, students' intrapersonal texts often contain a high quotient of discouraging views in anticipation of future interactions with teachers. Mindful writing makes this intrapersonal space accessible to student writers, turning over the keys to a room-of-their-own with walls made by breathing.

In this regard, freewriting and, maybe even more so, momentwriting afford individuals the opportunity to non-evaluatively capture their intrapersonal rhetoric on the page or screen, something otherwise seldom experienced. Due to the epigenetic nature of freewriting and momentwriting, these methods as recorded intrapersonal dialog let students look at their thinking before it is shared with others, mitigating the potential impact of assumptions and affective responses. This recording of intrapersonal is affirming and respectful, a holistic take on students as writers that prevents the sequestering of the personal from the academic, the emotional from the analytic. In this way, intrapersonal writing lets students *be*, transforming writing into an "ontological act" during which "we enact a sense of ourselves as beings in the world" (Yagelski 3). Additionally, observed inner talk provides a forum for students to organize the polyphony of internalized contesting social voices, ones that John Trimbur describes as "speaking for different systems of authority" (217). The writing down of intrapersonal rhetoric assembles disparate intertextual bits, creating a sense of voice or that one has "spoken." At this moment of writing, handling one's intrapersonal rhetoric entails optimal use of the working conditions of the present moment.

Intrapersonal as Illusion Maker

As an illusion maker, intrapersonal rhetoric leads us astray by turning our attention from present circumstances through absorbing hypothetical scenarios. Storylines are scripted by the discourse of the intrapersonal. They can be wild rides lasting several minutes in which we are absorbed in scenes of elsewhere, complete with a setting, event, and imaginary conversations, well stocked with assumptions. Frequently, those storylines are constructed from inflexible, repetitive thinking; not based on empirical evidence, storylines resort to the formulaic. As James Moffett says, we "ride on the same train of thoughts, and reach familiar conclusions. By means of redundant inner speech we maintain a whole world view of reality, a sort of sustained hallucination" ("Liberating" 305). Pema Chödrön describes three levels of this diversionary discursive thought, showing how a practitioner can avoid the temptations of storylines and transition from complete mindless to mindfulness. During the first level, "we're totally gone … Our thoughts take us far away from the present moment for a stretch of time. This is also referred to as fantasy. When you come back from these wandering thoughts, it's like walking into the room after having left it for a while; you've been somewhere else" (Chödrön *How* 63). With the second level, "You are maybe two or three sentences into a thought or story line, but you're not gone for long before you wake up and come back" (65). At the third level, intrapersonal rhetoric changes to cultivated internal rhetoric, a mix of storyline and thoughts directing the person to mindful awareness, or to "thoughts that don't draw you off at all" (66). As Chödrön describes this third stage, "You're sitting, and you put your mind on the object of breath, and you're staying with it; then there's this little vague conversation or in-and-out of thoughts that's happening on the side, but it doesn't draw you off" (66). Similarly, varying levels of consciousness are apparent during acts of writing; for instance, on occasions in which our own writing suddenly looks unfamiliar to us as we reread it, or as "language *emitted but not received*—or scarcely received" is imperceptible to us while freewriting (Elbow "In Defense" 275). (I once was surprised to "discover" a drawer of poems I'd written, believing that I hadn't done much writing that summer—not as crazy as it sounds.) Undoubtedly, the dominant storyline that writing students tell themselves features an audience for their still-to-be-finished texts.

In addition to its propensity for the illusionary, dealing with intrapersonal rhetoric is tricky because of its guise of non-consequentiality. Physicist David Bohm pointed out this trap regularly presented by thought: thought is highly effectual yet adopts a semblance of passivity. Bohm says, "thought is very active, but the process of thought thinks that it is doing nothing—that it is just telling you the way things are. Almost everything around us has been determined by thought—all the buildings, factories, farms, roads, schools, nations, science, technology, religion—whatever you care to mention" (10–11). The same applies to any composition: writing *never* happens without the intrapersonal, for the simple fact alone that the intrapersonal triggers the physical movement required to move a pen or airlift hands to a keyboard. Moreover, the intrapersonal is evasive: it projects discursivity without drawing attention to itself. Even in her investigations on mindful learning, Ellen Langer focuses on types of mindless thought (premature cognitive commitments, obeisance to rigid structures or categories, fixation on outcome) rather than the self-to-self rhetoric that engenders those types. In this handling of the intrapersonal, thought appears to be more a product and less a producer of effects and decision, less rhetorical. However, the real-time examination of passing internal thought that constitutes mindful metacognition is what allows us to understand and better manage those premature commitments and fixations.

Buddhism offers strategies to pay better attention to mental formations and the vehicles of the intrapersonal those mental formations ride in on to reach our consciousness. A high inner accountability based on critical thinking occurs in the Buddhist belief system. The opening of the Bible, "In the beginning, God created the heaven and the earth," juxtaposed with the oft-cited opening of a Buddhist scripture, "Mind precedes things, dominates them, creates them," highlights the Buddhist emphasis on metacognitive awareness (Thera 21). Buddhist doctrine establishes a metacognitive awareness from a trifecta of knowing, shaping, and freeing the mind (23). In writing, these steps are equivalent to observing, manipulating, and releasing ideas for content and ideas about process for the sake of exploration. Two Buddhist sutras, the Dvedhāvitakka Sutra or "Two Kinds of Thought" and the Vitakkasaṇthāna Sutra or "The Removal of Distracting Thoughts," address the consequences of thought and offer a procedure to mindfully perceive thinking. In "Two Kinds of Thought," a series of mental formations—sensual desire, ill-will, renunciation, non-cruelty—are managed as the meditator considers their effects on self and other, acknowledging how it "leads to my own affliction, to others' affliction, and to the affliction of both" (Ñāṇamolí and Bodhi 207). A subtle point is made that these thoughts can exert undue persuasion upon the rhetorical self, since "whatever a bhikkhu frequently thinks and ponders upon, that will become the inclination of his mind" (209). In reverse action, as the meditator contemplates each type of thought, the impact of thinking subsides for self and other because he "abandoned it, removed it, did away with it" (207). In a series of steps in "The Removal of Distracting Thoughts," meditators wrestle with their thoughts, told first to examine their thoughts, then to resist persistent negative thoughts, then to immobilize the thought-formation function of those thoughts and to "crush mind with mind" (213). The sutra offer picturesque similes to illustrate these abstract procedures, such as "Just as a man or a woman, young, youthful, and fond of ornaments, would be horrified, humiliated, and disgusted if the carcass of a snake or a dog or a human being were hung around his or her neck, so too … when a bhikku examines the danger in those thoughts, his mind becomes steadied internally" (212). It's here that mindful composing pedagogy deviates from Buddhist mindfulness practice by taking a functional, instrumentalist view on present temporality.

Mindful composition looks for a combination of directed and undirected thinking, a healthy balance between mindfulness and what would be called an inspired mindlessness. With a mindfulness approach to writing, we strive for clear awareness of our mental actions, trying to avoid outcomes of undirected thinking such as preconceptions and outcome fixation. Cognition can be categorized as "directed, undirected, and recurrent"; directed thinking works toward a specific outcome in a rational and organized fashion, undirected thinking wanders without a specific outcome, and recurrent thinking involves repetitive and sometimes unconscious thoughts (Kellogg 11). Ultimately, mindful composition operates with a goal that is antithetical to Buddhism: to help students reach a state of full absorption in a writing task with a possible outcome up ahead. First-year students practicing mindful composition train in how to notice intrapersonal rhetoric and mine it for its content and other resources—rather than observe and detach from it. A certain mindlessness is desirable, with the caveat that students periodically, if not often, return to mindful awareness to avoid becoming beached on inability to write. Practice in present-moment awareness is preparation for those moments in which flow ceases, as it invariably will cease. This pedagogy favors a mix of directed and undirected thinking, a cycling between mindfulness and mindlessness and back to mindfulness.

Mindful perception of intrapersonal rhetoric also identifies recurrent or habitual ways of thinking, those signs of mindlessness. Intrapersonal rhetoric, concocted from "re-runs, the self-haranguing, the re-cycled inner speech," can be prohibitive of creative-critical work because of its tendencies toward compulsion and redundancy (Moffett "Reading" 319). At issue is the way in which repetitive internal thought forecloses upon possibility by encouraging us to rely on established categories or single perspectives. Self-rhetoric can impede critical thinking whenever it causes rationalization or the justification of already formulated views, as Chaïm Perelman and Lucie Olbrechts-Tyteca explained (42–45). In Buddhism, this state of mind is described as "monkey mind" (sometimes the monkey is intoxicated), because the "random and unintentional movement of thought from one subject to another must be brought under control," or compared to an untamed elephant, because the "mind in its ordinary state is out of control" (Lopez 49). Similarly, James Moffett points out that the etymology of "discourse" is "running to and fro" ("Writing, Inner" 239). In the opening of a poem published in the *Buddhist Poetry Review*, David Evans alludes to monkey mind when the speaker pictures his mental state as "moving along uphill and downhill / with uphill and downhill thoughts / swinging wildly from branch to branch / in my head." The speaker finds himself automatically returning a greeting to a man in a pick-up truck, "unthinkingly at that!," only afterwards registering that the man is his next-door neighbor. This succinct poem shows the transition to gratitude (also the poem's title) that can happen when we catch ourselves standing in the midst of mindlessness.

Try lassoing your attention to a single-pointed focus—say the final sentence in the preceding paragraph—while watching your breathing. Your resolution to keep your mind on the sensations of breathing will be foiled by incoming commentary. What's likely is that you'll suddenly notice that you've been daydreaming, making plans for later that day, reviewing an earlier event, or resorting to rigid thinking and preconception. An experienced meditator at a month-long dharma retreat once remarked that he kept finding himself making elaborate plans for an expected inheritance of a set of late eighteenth-century furniture. Anyone who has practiced formal seated meditation knows that this internal talk is practically indestructible—it's why Zen practitioners call the intrapersonal "mind weeds." However, from a nondualist outlook, wandering monkey mind can be an asset, since the incredible range of monkey mind presents opportunity for novelty and the creation of categories encouraged by Ellen Langer. My wandering inner talk—if I observe it with mindful awareness, an important stipulation—can guide me to an assortment of usually diverse ideas in a short amount of time.

Māra in the Audience

Without a doubt, for writers, the most consequential illusion manufactured by intrapersonal rhetoric is that an audience is present during writing activity and that this apparent audience possesses immediate access to a writer's words as they're produced. It's the deep-seated misperception that writers occupy the same space at the same time as readers. The phantasmagorical character of audience is underscored by the fact that "the person to whom the writer addresses himself normally is not present at all" (Ong "Audience" 57). Somehow, as the first-year student works on a rhetorical analysis in his dorm room, his writing instructor perches on his roommate's desk, eats a microwaved Hot Pocket, and critiques the composition in progress. In actuality, any audience sensed during a present rhetorical situation is a construction of the student's intrapersonal rhetoric: an amalgamation of the writer's thoughts about the past and best guesses about an interpersonal future. Intrapersonal rhetoric is the self-to-self interior discourse that assigns a position inside the writing situation to an interlocutor self or a chimeric reader—often both as the experience fluctuates. Usually, much intrapersonal rhetoric is devoted to maintaining this illusion—more so than other rhetorical factors—because the commodity of the intrapersonal is words, just like the interpersonal. (Most of the time, we don't pretend that future observers perceive our current physical experiences, yet we behave as though future readers can perceive our words and thoughts well in advance.) For whatever reason, probably our education, the imagined reader who appears in our immediate rhetorical situation, visiting us *Christmas Carol*-like from the future or the past, doesn't seem to appreciate our developing draft. Instead, the imaginary creature always presumes the right to a polished text. The problem, of course, is that the text has yet to be finished—it, too, is an imaginary entity.

When interacting with illusionary non-present readers, student writers make a poor bargain, exchanging information that could be gained from their actual rhetorical context for best guesses, conjecture, and negative affect. An example I share with composition students was told to me by a yoga teacher friend. In this dharma urban legend, a man is driving late at night by himself during a blizzard outside Montreal when his car slides into a ditch. He sees a house in the distance with its lights on and decides to walk to the house for a shovel or assistance. As he approaches the house, the man begins thinking, "What if no one is home?" He answers his own question with "What sort of person would leave so many lights on if they're away?" He keeps walking and thinks, "What happens if the person who lives in the house is old or sick and unable to help me so I have to call the towing service?" He visualizes himself bothering the stranger if he has to wait in the house for the tow truck. He next asks, "What happens if someone is home but after looking through the peep hole they decide not to answer the door?" By the time he's arrived at the front porch, he's fuming, "What sort of a person wouldn't help a person on a night like tonight?" Through his intrapersonal rhetoric, he's told himself elaborate storylines which will likely affect his interactions with the actual inhabitant of the house.

The audience challenges presented by unnoticed intrapersonal rhetoric are symbolized in the legend of the Buddha as he resolved to sit meditating under a bodhi tree until he obtained enlightenment. On the third successive night, Gautama (his name prior to becoming Buddha) was taunted by the demon Māra, who was determined to keep him in the cycle of craving with "the last lash of

Ego" (Trungpa *Meditation* 16). The demon Māra is frequently called "the Bad One," yet he differs from a Judeo-Christian sense of the demonic in that Māra symbolizes the power of experiences to trap the mind rather than representing evil (Gethin *Foundations* 23). Riding in on an elephant, Māra first assaulted him with nine storms and then unsuccessfully with lust, thirst, and discontent, personified as the demon's attractive daughters (Lopez 39–40). Gautama was undeterred from his meditation. This part of the parable symbolizes the power of discursive thinking to lure individuals away from present-moment awareness. However, Māra's next strategy was to directly confront Gautama and "ask him by what right he sat there beneath the tree" (Gendlin 23). This vexation corresponds with a frequent struggle faced by students to view themselves as having the authority to write and enter a discourse community, typically scholarly. How Gautama chose to react to Māra is important, because he viewed the demon with non-violent loving-kindness rather than condescension, in part because he recognized that Māra was a projection of himself, a manifestation of his thinking (Trungpa *Meditation* 16). Gautama responds by touching the ground with his right hand—a gesture routinely depicted on statues of the Buddha—which then causes Māra to tumble off his elephant and his armies of distractions to flee (Gendlin 23–24). An analogous gesture for composition students is a placing of a "hand" on their immediate writing circumstance, claiming the cognitive-physical space for their own, banishing audience ghosts, and recognizing the discursive straying power of their own internal talk.

From a mindfulness perspective, audience is best considered as one type of arising mental formation, subject to the impermanence of the present moment. Composition scholars have described the phenomena of audience—addressed, invoked, evoked—in ways which highlight its nature as a cognitive phenomenon. For instance, Douglas B. Park hints at the made-up dimension of audience when he says that writers work with "conceptions of audience" and address "a basic image" of a group of readers by relying on an "idea of people-as-they-are-involved-in-a-rhetorical situation" (182–183). Park points out the assortment of verbs used to describe writers' actions concerning audience. Writers are said to "adjust to audiences or accommodate them [as well as] aiming at, assessing, defining, internalizing, construing, representing, imagining, characterizing, inventing, and evoking" them, all suggesting "imaginative dynamics" (182). Of noetic activities, imagination in particular is associated with audience. Peter Elbow directs us to the illusionary nature of audience when he states, "When we write, however, all audiences are in the head, even the real audience" (*Writing with Power* 187). According to Elbow, once writers reach this realization about their imaginary interlocutors, they are empowered and can learn to mitigate the impact of audience (187). Stopping short of fully attributing audience to thought, Elbow later walks the idea back by suggesting a combination of "external and internal readers," both possessing a shaping influence on writing ("Closing" 106). Walter Ong accentuated the notion that audience is a product of the imagination by calling it an act of fiction; readers resemble characters in a story because they are given roles to play by writers (Ede and Lunsford "Audience" 60). Lisa Ede and Andrea Lunsford's well-known distinction between audience addressed and audience invoked, and their advocacy of a synthesis of these approaches, also negotiates the real/unreal line. With audience addressed, a reader is a "concrete reality" holding attitudes and motivations that the writer tries to ascertain; with audience invoked, the writer acknowledges the impossibility of determining those details about real readers and resorts to dropping clues inside the text as to roles that are expected of readers ("Audience" 78–83). Subsequently, Ede and Lunsford amend this account to include conflict, disagreement, and failures to communicate to avoid glossing over the complexity of human interaction ("Representing Audience"). Yet even with audience addressed, which presumes the actuality of readers, what is contemplated by the writer is again a mental formation, since writers "rely on past experience in addressing audience to guide their writing, or they engage a representative of that audience in the writing process" ("Audience" 89). The line of inquiry that threads all of these discussions concerns where a reader "exists": either external to the rhetorical situation or internal and inside the text, residing in it as a series of signals: as Park says, either "outside the text" or "back into the text" views (184). Some takes on audience are also more or less literal than others, or more or less figurative.

As an apparition, intrapersonal rhetoric turned-into-the-shape-of-a reader is fleeting, flickering; as a mental formation, it is subject to the impermanence of the moment. This internal rhetoric is not necessarily a tidy Q & A between the writer's self and an imaginary interlocutor who perches on an imaginary stool facing the writer. Rather, this creation is usually more nebulous, shape-shifting, and hazy. Internal rhetoric happens between one part of the self and another without regard to the source of those parts: from the unconscious, reason, emotions, and so forth (Nienkamp x). Occasionally, the two parts of the self are twins, the interlocutor usually identifiable as the person's primary self and the part cast as the listener a close version thereof. For instance, Donald Murray describes an "other self" in which "the act of writing might be described as a conversation between two workmen muttering to each other at the workbench. The self speaks, the other self listens and responds. The self proposes, the other self considers. The self makes, the other self evaluates. The two selves collaborate" ("The Essential" 87). This inner self comes with roles and responsibilities to protect the writer's privacy, set up a productive work environment, and provide collegial support in addition to writing down material (90). In this explanation, the two parts are both cooperative "workmen," but in other cases, the "other" might appear differentiated from the self, the outline of a separate person. Regardless of the friendliness or similarity of this reader, these interior companions are fashioned from the intrapersonal, a point omitted in composing theory and pedagogy. Moreover, intrapersonally constructed audience doesn't really resemble fiction in that it lacks an intact story; many details are omitted and a presence intuited among the gappy consciousness of the internal. In certain moments, the audience is a composite creature, a combination of a few expected reader types, while at other moments, it's more a trace of a reader in the rhetorical situation. As ever-changing mental formations, audience entities resist neat categorizations of "dangerous" or "safe."

A mindfulness perspective calls for a closer, more fine-tuned examination of these illusions to practice detachment and not succumb to elaborate storylines about writing ability or writing task. As James Moffett said, "only when the individual brings some consciousness to the monitoring of the stream of experience does she start to become the master instead of the dupe of that awesome symbolic apparatus" ("Writing" 235). For one thing, audience formations will continue to arise as part of our intrapersonal rhetoric. There's no one-time fix or inoculation against the persuasiveness of internal rhetoric except an ongoing effort toward mindful awareness of the rhetorical moment. In Buddhist parables, Māra returns to pester the Buddha on several occasions after his enlightenment under the bodhi tree. Māra haunts the Buddha outside of monasteries, as the Buddha is recovering from a foot injury, while the Buddha is teaching, sometimes shape-shifting, like the time Māra transformed into a disheveled farmer in search of his oxen. Each time, Māra tempts the Buddha to an abuse of power, debates his meaning, or taunts his handling of physical experiences such as pain (Bhikku "Kassaka"; "Sakalika"; "Nandana"; "Rajja"). Each time, Māra leaves defeated, "sad & dejected at realizing, 'The Blessed One knows me; the One Well-Gone knows me'" and vanishes, much as an observed passing thought, once recognized and "known" as such, vanishes.

A mindful perspective on the tricks of the intrapersonal, including audience apparitions, calls for an alert but even-handed approach. Vigilance is required because of the seriousness of the impact of unobserved monkey mind, and Māra's potential for harm is exemplified when the demon triggers the Buddha's death. Reminding the Buddha of his earlier promise to transition to nirvana once he had finished teaching the dharma, Māra is said to have instigated the Buddha's decision to knowingly eat from a plate of food that would cause a fatal condition of dysentery (in other versions, it's a poisonous mushroom or truffle). That the Buddha knew the meal was lethal is said to be evident in how he had ordered the meal to be buried to prevent others from consuming it (Lopez 55–56). Buddhist practice asks practitioners who are mindfully aware of thought not to scorn seemingly negative mental formations, a generosity paralleled in the layperson ritual of offering dough images of Māra gifts and food in order to control such demons "by bringing them into a social relation, the position of the guest, to be offered hospitality" (Lopez 192). Meditation practitioners are taught a light approach to inner discursivity: simply to label it "thinking" and move on to the next breath without a big parade of self-recrimination (Chödrön *When* 20). It takes a "dispassionate and brief form of mere 'registering'" that is often more efficacious than using cognitive stages such as a "mustering of will, emotion, or reason," moves which might backfire and lead to a stronger demon and storyline (Thera 42). For the purposes of writing, a similarly gentle but steady awareness of audience is advisable. At the end of "Closing My Eyes as I Speak," Peter Elbow offers a pragmatic list of ways to monitor inner audiences by asking oneself honest questions about what's really happening in one's writing experience (110–111). It would be important for students to learn how to gently conduct these inquiries without becoming caught up in reactive storylines, a distinct possibility once an inner audience steps forward and makes itself known.

The No-Self of Intrapersonal Rhetoric

Buddhist theory adjusts common misperceptions of selfhood, clearing away dualistic understandings of self by offering a middle path between complete belief in an eternal self and annihilation. Along with the concept of void, the principle of *anattā* stands at the very core of the Satipaṭṭhāna Sutra and Buddhist mindful awareness and distinguishes it from other religions (Fink 289; Thera 75). Its centrality is reflected in how fifty-four of sixty-two common human perceptions covered by the seminal Brahmajala Sutra or "The All-Embracing Net of Views" focus on the self (Gethin *Foundations* 148). Interestingly, in Buddhist theory, "self" encompasses all phenomena, not just human, and refers to "[a]nything that exists autonomously, independently, or objectively" (Lopez 29). Differences on selfhood divide the two major factions of Buddhism, Hināyāna and Mahāyāna, with the former applying no-self to the fullest extent, negating self and permanence, and the latter replacing notions of self with interconnectedness (Watts *The Spirit* 27). This scope speaks to the egolessness of the Buddhist approach, since everything is on a level ontological playing field, further dissolving artificial distinctions. The false view of enduring selfhood results either from illogically maintaining the notion of an abiding, unchanging self or from behaviors of craving and attachment (Gethin *Foundations* 145). Both missteps lead to human suffering: the first through logical inconsistencies the believer will inevitably run up against and the second through the choices and behaviors affiliated with greed and defensiveness. In proposing *anattā or* no-self, the Buddha countered popular perceptions in sixth-century BCE Brahmin culture in India, documented in the Upaniṣads, of "an unchanging and constant self that somehow underlies and is the basis for the variety of changing experiences" (Gethin *Foundations* 134). Anyone who maintains the illusion of a constant self shortly meets up with several contradictions: they are unable to control the five aggregates routinely associated with personhood (body, emotions, ability to identify objects, conditioning factors, and consciousness); they can't reconcile the painful or sorrowful experiences of the five aggregates with an impermanent self; and the very notion of a self becomes meaningless without consideration of its particular but also fleeting experiences (136–137). As the Buddha said, "that which is transient is subject to suffering; and that of which is transient and subject to suffering and change, one cannot rightly say:—This belongs to me; this am I; this is my Ego" (Goddard 27). The second false view of selfhood is more mundane; once individuals believe they possess unique characteristics definitive to themselves, they covet those characteristics, fighting in great and small ways with their fellow humans and wreaking havoc.

No-self doctrine replaces unhelpful views on the self with ones that take into account the impermanence and potential for connection to be found in internal talk. The negation of self for which Buddhism is renowned is not a blanket negation but, instead, a specific repudiation of the characterization of self as autonomous and enduring (Gethin *Foundations* 145; Lopez 29–30). What remains, then, for self if stripped of these two qualities? What is left without the binary of a permanent or completely non-existent self? The Buddhist answer is that what remains is the middle way—or the ever-changing passage of consciousness, lifted from under the thumbtacks of a self, a "conceiving of persons in terms of sequences of causally connected physical and mental events rather than enduring substances" (Gethin *Foundations* 157). Through bare attention, a practitioner observes the arising phenomena of the present moment as impersonal and "the seemingly uniform act of perception will, with increasing clarity, appear as a sequence of numerous and differentiated single phases, following each other in quick succession" (Thera 36). Liberated from a heavy sense of self, this observed internal discursivity is only possible because of ongoing dependent origination in which all things depend on other things for their existence, a phenomenon of causality and interconnection. No entity is self-created and independent, uninflected by other entities, and to perceive the intrapersonal is to perceive that connected dynamic, the you in the I and the I in the you. In an interview at *Tricycle*, Thich Nhat Hanh called this interconnectivity "interbeing" or the "awareness that what we call 'I' is composed of 'non-I elements.'" This view of self can be useful in discussions of identity, such as Keya Maitra's application of an interconnected self to feminism-without-borders; here no-self theory "provides a theoretical underpinning for the moral development needed in the quest for liberation, namely, '*inculcating a life of selfless altruism*' not in the sense of self-sacrificing but in the sense of a self truly informed by its interrelationalities and interdependencies" (364). The porousness of this view of self has the potential to radically dissolve self/social distinctions, alter our sense of what constitutes ourselves, and reshape our interactions with others.

Paying mindful attention to the self can help us notice how the self is not solid but, instead, composed of countless aggregates and contingencies. Buddhist no-self doctrine divided human beings into five aggregates—form (organs as well as perceptions); feelings; ability to differentiate between objects; mental formations; and consciousness. Lopez says these "five aggregates are the inventory of what we call the person" and that they are "impermanent, none lasting more than an instant" (26). Due to the impermanence of self, "One can search exhaustively through all the aggregates, and one will not be able to find anything that does not disintegrate the moment after it comes into existence … What we call the person is simply a process, a chain of causes and effects, driven by the engine of karma" (Lopez 26). In a two-part meditation, Kathleen McDonald dissolves and reassembles our notion of ourselves by asking us to contemplate our physical and mental disintegration into "atoms separating and floating apart" with "[b]illions and billions of minute particles scatter[ing] throughout space," followed by a dispersal of our emotions and thoughts. Halfway through the meditation, she tells us to "check your feeling of I—where is it? What is it?" (56). The next step switches to the dependence of our bodies on organs, which are in turn dependent on cells and particles, and our minds as dependent on earlier experiences. She concludes that the "mere absence of such an inherently-existing I is the emptiness of the self" (57). It's precisely this perception of co-dependent arising that lets the intrapersonal serve as a path toward mindful awareness: without noticing this connectedness, our minds dead-end consciousness into static views.

Applied to writing, Buddhist approaches to selfhood help ensure that the intrapersonal remains a vehicle for mindfulness and that it does not perpetuate writing-related suffering. Specifically, Buddhist ideas about the self jettison any notion that there's a fixed self transmitting that intrapersonal rhetoric. Dwelling on the self (securing its happiness, adding impressions of stability, defending its "objects" or "possessions," including bodily attributes, from other people or from the ravages of time) detracts from attention to passing internal thought and therefore hinders mindfulness and mindful writing. Instead of confirming the autonomy of the self, in mindful writing theory, intrapersonal rhetoric is an indicator of the intertextual, interdependent nature of consciousness and composition. At the same time, it would be erroneous to nihilistically conclude that there is no individual who writes—only a broad width of social influence and structures. In first-year composition courses, students who mistake the self as permanent risk weakening their creative-critical thinking, losing opportunities for novelty, finding multiple possibilities, and divergent viewpoints. Conversely, students who are trained in a denial of the writing self stand in greater intellectual jeopardy, since what they are essentially taught to do is a wholesale erasure of the present rhetorical situation. How we manage intrapersonal rhetoric is linked to how we understand the writing self.

Static representations of self or mind currently found in composition theory can be redressed with the Buddhist view of no-self. Self and other are positioned as binaries in the writing act and even oppositionally, since there's inevitably a judgment involved. In this binary, to posit a writing self is to be individualistic, capitalistic, naïve, solipsistic, brainwashed, privileged, irresponsible, duplicitous, or corrupt; either that, or to suggest that writing as a social act is to be brainwashing, controlling, disrespectful, disempowering, uncreative, and inauthentic. It's hard to tell which came first, the static perceptions or the binaries about writing and writers, but they seem mutually reinforcing. The more we consider the writing self, for instance, to be autonomous and freestanding, the less likely we see it as subject to change, since change happens in collaboration with other entities and forces. To heap on troubles, the more a writing self is seen as discrete and unchanging, the more susceptible that individual is to writing problems and even writing blocks. This is because such a writer, through her self-casting, cuts herself off from a ready supply of change—from the shifting content for writing provided by shifting internal talk and from the shifting perceptions of writing self-efficacy also provided by that shifting internal moment. Such a student writer struggles to compose, not having anything seemingly to say, and perceives herself as locked into a singular ability level—*I'm a bad academic writer, I do a solid job with body paragraphs but never introductions, I can't write personally/analytically*, and so on. What results is reduced opportunity to develop a rhetorical self and work with one's inner rhetoric—both of which should be central learning outcomes in a first-year composition course.

Due to static notions of self and other, writing instead becomes a positivistic affair for students, an exchange of pre-set ideas, already provided by society, or more specifically, by previous academics and their texts. This line of thinking is the crux of David Bartholomae's argument with Peter Elbow and is a great set-up for mindlessness. This handicap is true even of so-called expressivist or subjective rhetorics, due to the intertextual mixture of the intrapersonal, because whenever we limit access to intrapersonal rhetoric, we are reducing social interaction. In composition theory and pedagogy, sometimes this dualism is a matter of agency and at other times it's about location. For instance, discourse becomes personified in the social constructivist view almost out of a compulsion to insert *some action* into a scene risking inertness sans student writer. For example, Nathan Crick says that David Bartholomae "effectively *personifies* discourse as if it was a self-conscious individual carrying out its desires by controlling actions of unknowing students. As his essay describes, discourse *determines* … discourse *works* … discourse *presses*, and ultimately discourse is *imitated*" (264). In Chapter 5 of this book, I describe contortionist performances made by composition scholars as they try to mitigate the consequences of their denial of writing self.

Furthermore, the location of self and the social frequently picks up a second dualism of interior and exterior, in which the interior is depicted as a place to exit, such as with Piaget's developmental model (Trimbur 215). This narrow view of where the self's activities take place results from a failure to distinguish between interiority and inwardness, in Stephen Toulmin's terms. According to Toulmin, individuals position writing and thinking as a matter of interiority because of the anatomical position of the brain and the nervous system, causing a sort of lock down inside oneself. However, in reality, writing and thinking are a matter of inwardness, which involves the purposeful and ongoing decision to bring notions and discourse from society inside through internalization (5–6). Encounters with the inward show it to be in a perpetual state of change, unlike confinement in the skull: "The things that mark so many of our thoughts, wishes, and feelings as inner or inward are not permanent, inescapable, lifelong characteristics. On the contrary, inwardness is in many respects an *acquired* feature of our experience, a *product,* in part, of cultural history but in part also of individual development" (5).

In combination, notions about the agency and location of the writing self will have implications for how the movement of thought and words is construed. As an example, if one holds too static a view of self and other, one might favor a model of communication that is bi-directional, involving a sharing of meaning-making between writer and audience: a cartoon 1–2 back and forth. A more mindful perspective on the writing situation, however, would suggest that this view is too simplistic, omits actualities, and stays cognizant of how language "is already and simultaneously inside and outside its users" (Trimbur 219). I may have learned a new word on a certain date after reading a *New York Times* article, and two months later a thought arises that I could use that word in a piece of writing, but the journey between those two moments is far more circuitous, dipping as well into blank period of unconsciousness, than a binary would suggest. Unclasped from binaries of self and other, interior and exterior, writing consciousness can become an activity of focusing on arising and falling mental formations and sensations, on the divergent and multi-inflected offerings of the moment. It focuses more on the moment-by-moment writing experience and less on the piece of writing, fostering writing as a way of being, at once a way of knowing the self and interacting with the ideas and words of others, Yagelski's writing as ontology. Take self away: watch changing mental formations. Any mental formation or bit of language emerging in the writer's consciousness, no matter how apparently socially sculpted, is included in the intrapersonal without sorting into categories of "me" and "not mine."

Ego-less writing is a polyphony; the intrapersonal is a monolog that incorporates multiple social elements. As a polyphony, the rhetorical self is "not a unitary self but a collective self," a proverbial crowd of sometimes contesting voices and information (Nienkamp 130; Trimbur 217). To further dismantle the self/social binary, differences between bits of internal rhetoric and social instances of language should not be overestimated, since "internal rhetoric involves interiorized social voices similar to those that shape external rhetoric" (Nienkamp 127). The diverse content of intrapersonal rhetoric—its polyphonic, intertextual complexity—is a result of the constant shifts in voices, outlooks, tone. That is, the flux of the intrapersonal happens because of a changing of the guard, from socially derived voices and ideas taking turns in the spotlight of the mind. What this makes possible is a connected knowing which takes what is known, enhances it, and creates new directions, instead of a separated knowing and disjuncture (Wenger *Yoga Minds* 25). In the views of some theorists, what organizes this phenomenological perception of the parade of the intrapersonal is a sort of managerial self, a part of the self that selects, interiorizes, and puts into relation (Nienkamp; Toulmin; Trimbur). Taken a step farther, even this residual self might be reconsidered from a Buddhist mindfulness perspective—whether there can be a "substantial self existing over and above the flow of consciousness," one which has a vantage point over events, or whether consciousness is a "purely experiential self" (Fink 239; 304). Fink doesn't even suppose that this "perspectival" self chimes with Buddhist no-self theory: "In reality, there is no deep sense in which I exist, and hence no deep difference between us, because reality is not perspectival" (304). Whether we adopt this radical sense of no-self or a more pragmatic middle view for writing, we are headed toward a mindful approach to writing. That is, as long as we remember to remember that intrapersonal rhetoric occurs in a silent individual, whose hand hovers around the keyboard, who interacts *intrapersonally and not interpersonally* during the present moment of writing.

In Right Audience, the mindful writer waits for internal language to arise in response to the large emptiness of the moment. Free of context, such words are often enigmatic, metaphoric, mixing with the impulses of the unconscious, as though words are large fish drawn to the surface, attracted by the opportunity of mindfulness. Such words and fragments emerge in response to the call of awareness, the reminder to remember the moment. Or the mindful writer steers his discursive thought, training the mind to generate words by asking a question, tossing the question into waves of breathing—questions about content, structure, an image, a single word, a comparison, a topic sentence, a thesis, a supporting sentence, an assumption, a rebuttal, a definition, a noun, verb, adjective, adverb, about a subject, a predicate, about punctuation, about a list, a paragraph, a line, a sentence. The mindful writer watching her in-breath asks a question of the moment, and the mindful writer watching her out-breath waits for the moment's answer. The mindful writer doesn't pursue language, doesn't fish in her discursivity, is not attached to produced words, doesn't sort inner words, but lets go of them like a catch and release, recording the words but suspending judgment. Stock your voice as you would an ornamental fish pond with phrases.

Interchapter 2

The Final Tally Was 1,390 to 744

The final tally was 1,390 to 744, two figures I wrote on the wipe board, a difference of 47%, as a mathematically inclined class member pointed out. The first number refers to the cumulative word count of the class for a two-minute freewrite to a writing mantra ("Your ability to write is always present") completed as a private writing. The second number represents the class cumulative word count for a two-minute freewrite to a quote from Suzuki ("In the beginner's mind, there are many possibilities; in the expert's mind there are few"), this time in a potentially shared freewrite. I announced beforehand that two students' names would be randomly drawn from the attendance roster after the activity and that those two writers would read us their second freewrite in its entirety. (Occasionally, I've used an intermediary freewrite in which I announced ahead of time that two randomly selected student writers would paraphrase their freewrite, or I've included a disposable freewrite in the mix.) I don't actually make anyone paraphrase or involuntarily share, but I don't disclose this plan until after the timed freewrites.

Along with the impact on word count, I document students' affective and embodied responses to the tasks: while they freewrote, I was taking notes on changes in posture (how close they positioned themselves to the notebook or computer screen); the pace and volume of typing (pockets of silence followed by staccato bursts or more timorous typing); and utterances such as sighs. I hone in on Taylor's typing and hear a different rhythm, the first freewrite more even in volume and speed but the second occurring in bursts and with some markedly heavy key-hitting. Some students have completely stopped freewriting with the second prompt and are staring at the monitor. Phoebe's typing remains continuous but slows down significantly for the second freewrite. Elizabeth's seated posture for the first freewrite had been perfectly straight and poised, but with the second she is leaning forward, toward the computer. A student writer in the back row of computers sighs three times in the space of half a minute during the second freewrite; another person holds her head in her hands, and since the set-up of the computer lab gives me a view of their screens, it's easy for me to also see that the second activity elicits far less writing.

The concept that audience is a matter of proximity, and specifically that the majority of audience situations are actually intrapersonal rhetorical acts, can be abstract for student writers, but "Summoning Audience Ghosts" provides them with quantifiable evidence. The numbers offer tangible proof of the impact of self-generated conversations on writing: their audience ghosts had been summoned and made visible. What's going on? Both experiences came out of a two-minute freewrite, a supposedly similar writing task. Afterwards, students spoke of feeling more cautious, of plotting out word choice and organization, of editing while composing. It becomes apparent to the class that their anticipation of the evaluation of an audience influences their word production as it happens, affecting both the quantity and their level of ease, and that, moreover, it's their own self-talk about that anticipated audience—and not the audience itself—that causes those changes at the moment of writing. Because, in fact, the moment of sharing never actually comes: as with any writing present moment, it can't be entirely predicted or known ahead of time. At the moment of writing, they didn't actually interact with a reader, only with their own self-talk about what it might be like to share the as-of-not-yet-completed freewrite with another person. For most students, this probably goes a step further into non-productive mindlessness, in that it actually *seems* like they're communicating with someone while they write; that self-talk is so vivid that it seems as if its posited future scenario is actually occurring.

What I most want student writers to comprehend about intrapersonal rhetoric is that it's immediate, quickly arising, and laden with changing content and approach, that "talking to oneself" is the basis of *all* writing, and that ignoring this fact leaves writers susceptible to writing difficulties as well as vulnerable to highly persuasive messages broadcast by internal rhetorics. The mixed blessing of monkey mind needs to become evident: that the intrapersonal is both a potential content provider and a carrier for preconceptions, distracting storylines, and other less beneficial matters of self-rhetoric pertaining to writing, including the manufacture of the figure of a reader. Another point that student writers need to understand is how working with intrapersonal rhetoric entails perceiving it, making what otherwise is inaudible and invisible a matter of observation, as well as learning to steer it, purposefully working with intrapersonal rhetoric as one might with more interpersonal forms of rhetoric, by cutting off the power supply to non-beneficial illusions such as audience proximity and keeping in check rhetorical assumptions about the writing-related future.

Before learning to monitor the intrapersonal for its mix of beneficial and disadvantageous rhetorical factors, student writers need practice in simply observing the intrapersonal for the production of content. Freewriting and *momentwriting* (discussed in Interchapter 4) are effective ways of capturing texts which are distinctly internal, and both freewriting and momentwriting can be harnessed to invention strategies like Peter Elbow's open-ended method, Sondra Perl's felt sense, and yoga for hands. Through necessarily low-stakes and often private tasks, students have the opportunity to engage with the fountain of inner discursivity that occurs in their minds and are unaffected by others' needs for clarity, etiquette, or interest. I emphasize two gestures—*stocking and steering*—to foreground the intertextual, social nature of inner discourse and indicate students' rhetorical control over it. Whether the intrapersonal is observed or generated by the writer is irrelevant, and the level of agency is really an ego question; however, it is the case that the intrapersonal can be steered or guided in a certain direction. As we observe our inherently discursive monkey mind, we invariably notice its broad-ranging, diverse, scattered, imitative characteristics: one second, we're hearing a possible topic sentence, the next something a roommate or colleague said or a song lyric. The good news about this permeability between self and the social is that there are boundless opportunities for engaging the ideas of others, for trying out different conventions, approaches to voice, language preferences. We can "stock" our intrapersonal as someone might stock a pond with fish, selecting through what we read the types of languages that infuse our internal texts. A writer might seek out the metaphoric by reading Emily Dickinson or academic rhetoric by looking for turns of phrases commonly used in scholarly works from specific disciplines. Exercises in code switching and code meshing also help students move between different qualities of intrapersonal voice in a single document. Writers can steer the intrapersonal by setting an intention or consistently asking themselves prompts while writing. With the intention in mind, writers silently ask themselves, What am I thinking about *x* right now? And now? And now? The key is to remain receptive to whatever material arises but then to keep up the inner Q & A, keep throwing the stick. The discursive mind will in all likelihood chase after it.

Several exercises rely on personification, similar to the memorable Buddhist concept of monkey mind, to heighten student writers' awareness of audience as by-product of the intrapersonal. In "Caricature of a Difficult Audience," students explore the frequently hyperbolic and distorted perceptions of audience that result from our intrapersonal rhetoric by creating a verbal caricature—similar to the type of caricature drawing that can be bought at a county fair or in a shopping mall—of one of their challenging audiences. This low-stakes and unrevised activity highlights how much imaginative work we put into our fake conversations. I give a series of in-class prompts to which they jot down notes for later development into a paragraph. First, I ask them to think of an occasion in which they were struggling to write: perhaps their writing was slowing down, but not in a calm, reflective way. They visualize someone who has an opinion about this piece of writing manifesting in the room or area where the student is trying to write. Building the caricature, they describe the person's face (making one or more facial features odd and the head disproportionately large, like a caricature drawing) and the person's body and clothing (making the clothing absurd in fashion, size, or tidiness). Phoebe selected a high-stakes task familiar to her peers, a college application essay, a difficult writing situation amplified by the anonymity of its evaluators. Phoebe gave her faceless reader the visage of a forty-five-year-old woman with "crow's feet that stretched for miles" and a lacey dress which "looked like it would be on your grandmother's dinner table." This reader "acted like a dictator; her word was final; nothing ever could change once she hated your essay." Likewise, Allison's genre was also a college application essay, and her tricky audience was a middle-school teacher who looked as though "he was drowning in his clothes" with a tie "so tiny it was almost invisible." For Amanda, the tricky reader is a high-school teacher wearing maternity clothes, though not pregnant, so that she "looks like a little girl playing dress-up," a teacher who "made me feel both a dangerous audience and a dangerous nonaudience." Jonah opted for the high-stakes situation of a state standardized writing assessment given to all high-school students. He gave a form to this unknown reader, a "random person" who appears at his desk wearing Crocs and immediately critiques Jonah's opening paragraph with the intent "to make me feel like not writing anything." Jonah's tormenter is even so duplicitous as to arbitrarily alter the exam prompt while Jonah is working so that the tormenter has less to evaluate.

With the next prompt, they imagine the person starting to say something about their writing and come up with the first sentence, immediately deducing that the tricky reader has an ulterior motive and explaining what it is. Visualizing a tiny difficult audience perched on their own difficult audience's shoulder, they imagine how the person was bothered in the past by dealings with a complicated audience—a nun (Taylor) or a harsh father (Amanda). Gradually, the primary tricky audience's speech starts to become distorted and out of their control, which frustrates the tricky audience. The student writer describes his or her next actions. Elizabeth describes how "When my tricky audience starts to talk to me, he ends up saying gibberish language that I don't understand, and he is probably the only person who understands, 'You AUE tGE MQsT DUMV PERSON EVEEEERRRRR.'" Amanda's critic tells her she'll never be good enough for college but regresses into baby babble to talk with that miniature figure of his father that's perched on his shoulder. Lastly, channeling Ann Lamott, students put their tricky audience in a container (any sort is fine—an old beer bottle and a litter box have been greatest hits) with an execrable substance at the bottom, and the container is disposed of in some way. Claudia puts her inner critic in a Tupperware container and moves that container "outside to the woods behind my house. This is what I do most times if I ever catch a mouse in my house. I know the woods is far away, and I'm sure that Mr. Z won't be returning to my workspace."

At the next class session, students complete a private freewrite in which they explore what the details and imagery of their caricature might symbolize about their relationship to this reader. For instance, could the fact that her nose turns red-hot like a stove burner be indicative of worry about the reader's anger? (When I did a caricature assignment with my students, I'd been struggling with a commissioned article on my writing mentor, a living American poet, and one reason for my faltering was worry that she'd be angered by my account of her oeuvre.) Could the fact that he wears a cardigan covered in boarding school and Ivy League insignia be indicative of feeling intimidated by the reader's higher social class and educational standing? Or the way in which the caricatured reader keeps pushing the student to use a bigger vocabulary the reader's own fear that she wasn't taken seriously? After discussing the symbolic nature of the caricatured audience, we discuss how it's fundamentally a fictional creation of the student. Tatiana, discussing the intimidation she felt from her audience, realizes he is "an insignificant part of my imagination that I gave power to." She says, "I wish that I had never created him, but it's almost instinctive to imagine someone who is always looking over your shoulder constantly critiquing your writing." Whenever students find themselves hesitating or feeling unsure while writing, they're encouraged to figure out who exactly they've imaginatively installed in their writing space, to take a good look at the absurd creature and remind themselves that their writing situation is actually vacant, free and clear of such imaginary creatures.

Later in the semester, students work on a low-stakes exercise, "Visit from a Writing Demon," based on the Buddha's encounter with Māra, in which they develop an analogous story of challenges presented to them (usually applied to a specific, current project in the course) by their self-talk. Just as Māra was determined to keep the Buddha ensnared in the cycle of craving with "the last lash of Ego" by assaulting him with distractions and challenges to his authority, students face similar vexations and distractions. Māra's final ploy corresponds to the problem faced by student writers of how to view themselves as possessing the authority to write. The Buddha's reaction to the demon—touching the ground with his right hand, a gesture routinely depicted on statues—causes Māra to fall off his elephant and his armies of distractions to bolt. An analogous gesture for writers is placing a "hand" on their immediate writing circumstance: claiming the writing moment for their own, banishing audience ghosts, and recognizing the discursive straying power of their own internal talk.

In the exercise, students describe a scene in which they return to work on a higher-stakes project in the course, a scene that can be literal or imaginative—sitting at their dorm desk or with their laptop under an ancient tree, for instance. They describe what they're doing in the draft the moment their personal Māra appears and imaginatively create that demon, giving it a name. Claudia describes a writing situation in which she is sitting in bed and trying to address edits emailed by her professor. A troll named Roy appears on her blanket, an occasional pest of her high-stakes work, and his interruptions escalate from a casual "Whatcha doing there?" to unplugging her headphones and taunting her that she'll never be a perfect writer. Explaining the damage caused by taking her internal negative talk about her writing too seriously, Amanda says "my Māra thrives off of attention and seeing me do badly." Phoebe's demon increases in size with every comment, reflective of the impact of pessimistic self-talk about writing, as it accrues and impinges upon future writing experiences.

Students next describe a series of moves the demon makes to distract or discourage them and their reactions to each of those demonic moves. I give this exercise later in the course because the expectation is that students' countermoves will involve the various strategies taught in the course for managing audience proximity, engaging in prewriting and invention, and maintaining awareness of the present rhetorical situation. Students describe returning to freewriting, practicing mindful breathing, using yoga for hands, switching up the imaginary audience in their head to one who is powerful and friendly, and even practicing empathy with this intrapersonally generated creature. Marcus' writing demon "comes through the ceiling along with the flames surrounding her. All of a sudden, the room is noisy, almost like I am sitting in New York City traffic" and starts her temptation process by inviting Marcus to play basketball or watch YouTube videos. Just trying to evade the demon only puts Marcus into a "dark black room" where he can't seem "to be able to think of complete thoughts to write," so he resorts to freewriting. Once Marcus freewrites, the demon shapeshifts into a crowd of "hundreds of people staring at me giving me dirty looks," forming a dangerous audience crowd, but through his mindful breathing, even these demons depart, leaving him calm enough to write. Allison's demon seems "magnetized to my keyboard" and then stands on her shoulder playing with her hair to distract her. Her solution to alter her intrapersonal negative talk is to begin singing very loudly out loud. Jenny's demon, named Stall, comes from a place of perfection which makes our actions and feelings seem out of our control—we're driven by an illusion. Malissa's demon goads her to write too fast, suggesting that someone is impatient with her and not respecting her natural writing rhythms. Interestingly, Malissa uses freewriting to gain distance from her demon—a strategy that tends to pick up the pace of our writing—banishing the demon. Tatiana is caught in the act of last-minute dangerous method writing by her demon when a "disfigured human with a clock for a face stands in my direction: no sound, only movement. I don't want to look at him, but I can't take my eyes away from the ticking hands on his face" and uses mindful breathing to "evaporate" the demon and give herself an "open mind." For Elizabeth, freewriting converts her demon, "My monster turned into my angel." Ideally, students should use several methods to counter the several attempts by their writing demon to deter them, and by story's end, an ultimate strategy should compel the demon to altogether vanish, leaving the student free to continue writing.

Another activity which directs attention to imagined audience involves considering how certain writing materials—composition notebooks, a Word document, Post-Its, handsome journals, magic markers, exam blue books—contain traces of audience dynamic and impose expectations on the writing moment. This activity doesn't resort to personification to capture the usually elusive audience-in-the-head but instead asks students to think of the ways the material conditions of writing suggest contexts with particular audience dynamics. What types of writing are normally done on Post-Its, for instance? If they began their piece on Post-Its, how might that association help or hinder their effort? Do they need to write the next draft on a blue book to pick up the pace by imagining a get-it-right-the-first-time timed writing scenario? We practice ways to manage audience proximity by selecting materials with greater distance from imaginary scenarios of already-written final products handed to imagined audiences (paper from recycling bin, crayons) or materials with closer proximity to those scenarios (greeting cards, writing essay in body of an email to the professor). This check-in can occur at any stage of a writing process—so materials might be selected at the start and then modified midway or toward the end in order to return to writing activity resembling prewriting, at one extreme, or to move to writing activity resembling editing, at the other. The idea here is that intrapersonal rhetoric doesn't just occur in the earliest phases of prewriting and invention but is ongoing, just as the present moment happens continuously throughout the experience of writing. The second benefit of considering the material conditions of writing is an enhanced overall awareness of the present writing moment that comes from attention to one of the objects of the moment. Essentially, these kinds of activities are asking student writers to identify ways in which intrapersonal rhetoric is distinct from interpersonal rhetoric. Chiefly, the inner text of the intrapersonal—with its moment-by-moment experiences, its mix of the fragmentary, the embodied, and even the nonverbal—is never identical to an interpersonal version shared with readers, a point further underscored by momentwriting.

Second only to its installment of illusionary audiences, one of the most consequential impacts of intrapersonal rhetoric for student writers is its provocation of preconceptions, either about general writing ability or task specific (often what will happen next, what will be covered, where difficulties will arise, what the final product will look like, etc.) In Buddhist mindfulness, preconceptions are storylines that drag us off our awareness of the present moment with their alluring or disturbing fantasies. Accordingly, in "Exaggerate a Storyline," students select a recent or reoccurring story they tell themselves about their writing ability and in 500–750 words follow it to its full-blown conclusion—a dramatic, overblown, and likely silly view that highlights the crazy fictional nature of some of the products of the writerly intrapersonal.

While many activities involving the observation of intrapersonal rhetoric for the purposes of identifying the generation of rhetorical factors are low stakes, two higher-stakes and revised assignments I use are a self-interview and an internal rhetorical analysis paper. The self-interview reinforces the inner Q & A and steers students' intrapersonal rhetoric toward greater awareness of their writing practices. I frequently bookend the course with these interviews, such that the first assignment in the course is a self-interview and a sequel functions as the equivalent of a final exam or portfolio. Students perform the role of both interviewer and interviewee, asking specific types of interview questions (closed, open, mirror, hypothetical, grand tour) about their writing experiences and then composing a response. The questions are designed to explore particular parts of the composing process, including mindful writing factors such as affect, voice, audience proximity, and preconception. An open question might be "What are a few good reasons for staying aware of the present moment while trying to write?" or "What are the most effective ways of managing the impact of audience?" A hypothetical question might be "What would you tell your younger high-school self about prewriting?" A grand tour question could be "Describe everything involved in remaining aware of the present moment when receiving feedback from another person on your draft." In explaining the reasons for present awareness for writing, Allison wrote:

> There are endless good reasons to stay aware of the present moment while trying to write, but one of the best is that it allows you to stay in your thoughts and your own consciousness. You are focused on yourself and your breath and body in that point in time. It stops you from worrying about the

> future of your writing, and it makes the whole writing process a step-by-step method for the best results.

To the same question, Jenny responded that a "good reason to stay in the present moment is you aren't passing judgment on yourself from that audience in your head because you're focused on the writing." For the sequel, students reflect on the difference that fifteen semester weeks have made by reviewing the earlier interview and incorporating incidents in the semester and excerpts from informal and formal writing assignments as evidence.

The internal rhetorical assignment is a mindful version of the conventional analysis of interpersonal rhetoric frequently taught in first-year writing courses, except that an internal rhetorical analysis does more to develop student writers' metacognition. Once student writers become conscious of their ongoing inner discursivity, they can begin to analyze and even refashion it in the spirit of the Dvedhāvitakka Sutra and the Vitakkasaṇthāna Sutra. It is striking that our discipline has offered so much in terms of how to analyze interpersonal rhetoric, and yet we have no examples to my knowledge of the systematic rhetorical investigation of internal rhetoric, one of the many points on which I agree with Jean Nienkamp. Lacking a systematic investigation, for instance, of self-ethos or the way in which the self represents itself and its abilities *to itself*, we are at the mercies of this invisible agent.

Internal rhetorical analysis involves close study of the connotative language, rhetorical appeals, and rhetorical assumptions of self-communication as well as the intended and unintended suasive effect on the self as audience. This type of assignment reinforces the perception of intrapersonal rhetoric in relation to other factors—such as impermanence—to prevent a static, reified understanding of the intrapersonal. While presumably the topic of this internal rhetoric could be anything (anything the self talks to itself about—dieting, relationships, career, plans, politics, etc.), for our purposes, the topic would be the act of writing. This topic can also be sub-divided into writing in general (so the internal rhetorical analysis would look at how the student self-discusses overall writing abilities and experiences) or writing specific to a task (so a particular rhetorical situation would be examined). The intent is to examine that primary, first-on-the-rhetorical scene text that happens when an individual writes, that blend of phrases pertaining to the content of the emerging draft and also of meta-commentary, asides, and evaluations. The text examined in an internal rhetorical analysis consists of any instance of discursivity that arises without differentiating between the privately spoken and what might be shown to an eventual reader. Normally, when we write, we say and hear these asides and commentaries about our writing without typing them (unless we are freewriting): they become fleeting flotsam not included in a written product. The point of examining this flotsam, however, is that it often exerts sizeable influence on the writing experience and on the written outcome.

Capturing this inner text for the purposes of analysis can happen through several methods. One method is to have student writers practice momentwriting as a draft of a piece of writing that will eventually be interpersonally shared. Another method is to make it a tandem writing task in which students set a sort of "alarm" or reminder that interrupts their otherwise mindless writing of another text and asks them to check in, pinpointing particular rhetorical factors as the focus of a reflection—for instance, their self-ethos at that moment or various assumptions they have made in the past few moments of working on the piece. *What are you persuading yourself to do or think concerning your writing? How are you talking to yourself about your writing in the past few minutes: what tone are you adopting? How much or little authority does your self-ethos exert over yourself and how is this established through your internal rhetoric? What sorts of emotional appeals are you using on yourself about this writing activity?* Rhetorical assumptions, really a type of preconceived notion, would be particularly interesting to analyze. *What are you assuming, tacitly or explicitly, about your ability to complete the writing task at hand or about the outcome of this writing task?* A third method entails selecting a single sentence of self-rhetoric and examining it for its assumptions. Still another method is to have students describe how they speak to themselves in general about their overall writing ability, scanning for tendencies which can be examined for their rhetorical construction. Finally, an assignment in internal rhetorical analysis involves building an altered internal rhetoric: a persuasive essay, as it were, addressed to the self with the purpose of convincing the self to a adopt a different stance toward a particular subject or toward writing (in general or task specific). Students manipulate the rhetoric to attempt to achieve a particular change in themselves and reflect on their rhetorical strategies.

Observing the intrapersonal is crucial to an ongoing maintenance of mindful awareness, and so attention to the intrapersonal must occur throughout the process of writing. Training in intrapersonal rhetorical awareness happens during prewriting, drafting, rewriting, receiving feedback, and editing as a way to constantly return to the present moment. For example, using the "Montaigne method" of invention, writers monitor the extent to which their intrapersonal rhetoric about writing acts is skewed toward negative commentary. This method has students invent purely through the addition of material by banishing the deleting, correcting, and editing that often become interlarded with generating (see Thomas Newkirk "Montaigne's Revisions" and *Slow Reading* as well as the article I co-authored with Staci Fleury). An example of intrapersonal rhetoric as the featured component during feedback is "No Feedback," in which writers solicit feedback *from themselves* in response to the real-time reading of their pieces by another student. One student silently reads another student's draft in front of him or her without saying a word. The writer of the piece, watching his or her inhalation and exhalation, observes this reading from a few inches away and takes private notes on his or her own internal conversation about being read in real time. The writer jots down whatever internal language passes through his or her mind on the occasion of being read—*I hope she appreciates that second paragraph. Did I forget that definition on page three? I wonder if I'm clear enough about my reasons.* It's a sort of "laugh track" but instead a feedback track, and instead of anticipating a future hypothetical audience, the student sees an actual flesh and blood audience sharing his or her present moment along with the draft. In my writing practices, I've noticed that I often discover a solution to a vexing structural problem a few minutes after I email a draft to a writing colleague, before hearing that person's suggestions. The close proximity of this audience and the real-time occurrence of being read amplifies the student's guesses about the reader's responses, but they're nevertheless still *guesses*, framed by the student's internal discussions. This activity highlights the intrapersonal nature of each moment of composing, how no matter what the stage, our writing is curated by our internal rhetoric. It can complement feedback activities in which the peer reader does disclose responses to the draft.

3

THE VERBAL EMPTINESS OF MINDFUL INVENTION

> Form is emptiness, emptiness is not different from form,
> neither is form different from emptiness, indeed, emptiness is form.
>
> —*The Heart Sutra*

Mindful invention means facing the present moment for the purposes of writing, and it means managing the paradoxical co-existence of the nonverbal and the verbal in each writing moment. From a mindfulness perspective, each moment in the string of thousands of moments that comprises a writing experience, no matter how far along the moment falls in the advancement of a draft, is an empty (preverbal, nonverbal) event. For one thing, the nonverbal initiates every single present rhetorical moment; the observed moment starts off as non-discursive with the embodied sensations of breathing and often turns discursive within split seconds. A fresh present moment opens with awareness directed to the expansion of the lungs, the air pushing against the ribs, the temperature of air as it moves past the nostrils and then shifts to bits of words or voice, half an image, or the tail-end of a sentence of intrapersonal rhetoric. As a result, a mindful writing moment during prewriting is dunked in nonwriting, and it's the same with a writing moment during the revision of a middle draft and a moment during the last proofreading decision. Second, with mindful invention, what is perceived are the instants when formlessness turns over to form, when out of observed emptiness emerges a phrase, and back again, when form turns over to formlessness and emptiness manifests after a stretch of writing. The primary difference between mindful invention, neoclassical invention, and the invention of process pedagogy is the emphasis of the first on an ongoing and productive rhetorical emptiness, which includes nonwriting and the nonverbal, and which is usually confined, if discussed at all, to the prewriting phases of composing.

This chapter explores the Buddhist concept of the mutuality of form and formlessness in *sūnyatā*, an emptiness in which it is said that all things, not just the human ego, lack independent existence. *Sūnyatā* is not the annihilation of existence but, rather, the repudiation of a particular kind of existence (independent and permanent), replaced in a mindfulness perspective with an interconnected and continuously changing one. Approximately 400 years after the Buddha's death, a series of texts were developed in India to instruct on emptiness and lead to *prajñāpāramitā* or the perfection of wisdom, the most renowned of which is the *Heart Sutra*, in which the Buddha described a perpetual interplay of form and formlessness. For first-year composition, the wisdom of the Heart Sutra resides with its potential for a nondualistic interplay of writing and nonwriting: writing emerges from nonwriting, and nonwriting emerges from writing, carrying important implications for how writing ability is construed. The *prajñāpāramitā* carries special insight into the radical contingency of knowing or *groundlessness*, and its premise can overturn the dualistic thinking that too easily obstructs writing. To overvalue one genre or stage of writing over another, and perhaps especially to overvalue form over formlessness, is to risk writing suffering, since in so doing a writer is craving a particular product and overlooking his or her present rhetorical situation. In Buddhist theory, several terms are used for the mindset needed to reflect on emptiness, with a distinction frequently drawn between concentration and awareness: *clear seeing or comprehension*, *bare attention*, *luminous mind*, *bodhichitta*, *Buddha Nature*, and *beginner's mind*. In this chapter, I explain the reasons for guiding student writers toward an understanding of verbal emptiness as well as methods derived from Buddhism for observing emptiness, including emptiness of self, as part of writing metacognition.

The idea of emptiness is important in at least two regards for first-year composition pedagogy and for invention theory: first, many writing students fail to observe emptiness and miss out on an important resource for generating writing; second, probably just as many students avoid the nonverbal and those times of not writing, misconstruing it as a sign of a writing block or of their general writing inability. Normally, not-writing is rejected experience. Students worry that it means they'll look lazy in the eyes of their teachers or that they lack the abilities required of a successful college writer. Meanwhile, teachers do everything to avoid not-writing, ushering students quickly to the verbal by providing kick-starter prompts and seldom including nonverbal components such as writing embodiment in the curriculum. Don Murray once criticized teacher-provided prompts, preferring a more student-directed prewriting, because he felt that teachers relied on exercises because they "often do not have enough faith in their students to feel that the students have anything to say" ("Writing as Process" 6). Although I don't entirely subscribe to Murray's take on teacher-provided starts, I do think it's preferable for student writers to approach invention with few provisions in order to better understand the complex silences of prewriting and avoid long-term writing blocks (Peary "Terrain"). Perhaps teachers also worry that if their students engage in not-writing it's a sign of their failure as teachers—after all, it's a writing course and not a blank page course. If prewriting activities are hard to grade, certainly students' non-verbalized writing activities will fall uncomfortably outside the scope of accountability. Instructors' compulsion to avoid nonwriting also stems from just wanting to alleviate the anxiety first-year students commonly feel about starting writing assignments. Writers at every professional level and in every genre, even career writers, associate not-writing with failure, provoking anxiety such that the fear of not writing may rule their practice.

Typically, first-year writers associate nonwriting with what happens when they start an assignment. Anecdotally, in my summertime review of over 1,100 writing samples of incoming first-year students for writing self-placement at my university, essays in which students make a case for either a non-required basic course or the required first-year course, many attribute their selection of the basic course to a perceived problem with invention during high school—that it takes time for them to start writing. *It takes me forever to start an essay after it's assigned—sometimes even a few hours. A teacher gives out the prompt for a project, and I'm so slow that I don't know what I want to say until a day later.* I am flabbergasted by students' stringent expectation that they should be able to immediately start writing—wanting to point out that even their teachers would need time to mull over a new project—and by the extent to which nonverbal experiences of writing are rejected. We should be concerned when student writers expect continuous verbal production to come out of each rhetorical moment. While a steady supply of discursivity is made possible by our intrapersonal rhetoric, such that anyone could begin writing by observing the rhetorical moment, even this capacity should not obviate the invaluable contributions of the not-writing and the nonverbal in the writing moment. Because most first-year writers associate nonwriting with what happens when they start an assignment, it's probably most practical to tie conversations about verbal emptiness with what's normally thought of as prewriting activity. Eventually, first-year writers will need to understand verbal emptiness as an ongoing component of mindful writing or as a phenomenon of each writing moment, no matter what the stage of development. The study of verbal emptiness results in a new sense of what a beginning is for a writing task. Each observed moment during writing constitutes a beginning, in part because each moment contains a dimension of nonverbal, preverbal elements. So there's no one specially demarcated area in the process of writing a document that's marked off (usually with a rope of stress) as the starting line.

The study of form and formlessness reassures student writers that a wordless stretch will turn over to words because of the nature of impermanence—a promise, if we can only rest with that wordlessness and cease denying, avoiding, or mislabeling it. Ultimately, the concept of emptiness deconstructs binaries of writing ability/inability and increases students' self-efficacy and confidence. Put simply, the paradox of emptiness is that nonwriting is included in every instance of writing, and writing is included in every instance of nonwriting in the work of a mindful composition course. In *Zen Mind, Beginner's Mind,* Shunryu Suzuki says, "When the Buddha comes, you will welcome him; when the devil comes, you will welcome him … There is no problem" (42–43). Similarly, when writing comes, we welcome it, and when not-writing comes, we also welcome it. We can perceive verbal emptiness in every breath and during every present writing moment—it's nothing unusual or worrisome.

The Heart Sutra

Dating between 350 and 700 BCE, or approximately 400 years after the Buddha's death, possibly composed in China, the Heart Sutra offers insight on the movement between form and formlessness with implications for writing. The Heart Sutra is said to be the shortest and most popular of the genre of sutra called the "perfection of wisdom" or *prajñāpāramitā*, a group of forty texts concerned with nondualism and emptiness. This genre of sutra guided individuals who aimed to become *bodhisattva*, or those who could achieve enlightenment but have delayed their exit from suffering in order to assist others, unlike the *arhat*, who is an individual who achieves nirvana and exits the world. This difference in agenda distinguishes the two major sects of Buddhism, the Hināyāna and Mahāyāna.

In the Heart Sutra, two disciples engage in a Platonic-style dialog, Sariputra asking Avalokitesvara about the best way to study for wisdom, while the Buddha, in an early example of teacher-decentered pedagogy, meditates nearby in *samadhi* (his mind perfectly still). Avalokitesvara responds, "Form is emptiness, emptiness is not different from form, neither is form different from emptiness, indeed, emptiness is form" (Goddard 85). Next, in the rhetorical style of traditional Buddhist texts, Avalokitesvara adds examples and categories, saying that "sensation is emptiness, emptiness is not different from sensation," repeating the equation for perception, discrimination, consciousness. It's due to the impermanence of all entities, including the self, that nirvana is actually a possibility. As Avalokitesvara explains, "Thus, O Sariputra, all things having the nature of emptiness have no beginning and ending … it is only because personality is made up of elements that pass away, that personality may attain Nirvana" (Goddard 85–86). What makes this teaching particularly mind-opening is the extent of its application, since emptiness applies to self. Prior to articulating this now famous mantra about form and formlessness, Avalokitesvara began by addressing the issue of selfhood: "If a son or daughter wishes to study the profound Prajna-paramita, he must first get rid of all ideas of egoselfness. Let him think thus: Personality? What is personality? Is it an enduring entity? Or is it made up of elements that pass away?" (Goddard 85). To top it off, the lesson of emptiness is so pervasive, persisting without beginning or end and applying to all matters, that it extends even to the very teachings of the Heart Sutra, impacting as it does all principles, all concepts, all teachings, even the Buddha's teachings.

By the sutra's end, emptiness of form also applies to the Heart Sutra, and to the rest of the Buddha's teachings. Astonishingly, it affects the seeming grand prize of mindfulness endeavor—enlightenment and suffering's end—since "[t]here is no Noble Four-fold Truths ... There is no knowledge of Nirvana, there is no obtaining of Nirvana, there is no not obtaining of Nirvana" (Goddard 86). Talk about antifoundationalism! It's a radical view of contingency that might be welcomed by post-process proponents who have decried any codification of writing process or what Gary A. Olson described as "a Theory of Writing, a series of generalizations about writing that supposedly hold true all or most of the time" (Breuch 130; Olsen 8). Apparently, "Form is emptiness and emptiness is form" so shocked monks attending the lecture with its call for epistemological flexibility that some suffered heart attacks on the spot, while other less dramatic accounts said that monks absquatulated in protest (Chödrön *The Places* 103). The sutra closes with a mantra, "*Gate, gate, paragate, parasamgate, bodhi, svaha!*," translated as "Gone, gone, gone to that other shore; safely passed to that other shore" (Goddard 86). As Judith Simmer-Brown points out, the exclamation point at the end of the mantra is "roughly equivalent to the Americanism Wow!" because each syllable of the mantra is intended to lend a visceral rather than logical experience of enlightenment (xxi).

This Buddhist view on emptiness refutes notions of the essential, fixed, or enduring in favor of a socially constructed and situated concept of dependent origination. Emptiness is "the Buddhist technical term for the lack of independent existence, inherent existence, or essence in things" (Garfield 88). Emptiness also denotes the dynamic between form and formlessness: both their fusion and their tandem operations. So while the initial connotation of emptiness might seem desolate—might seem to tilt toward the formlessness side of things—emptiness in this sense refers to our receptivity to nondualistic types of thinking and experience. Every entity is empty because it does not display qualities that would make it discrete, autonomous, or freestanding; instead, each entity is completely interconnected with other entities, sharing the qualities of those other entities. As Edo Shonin and his colleagues describe it, "one could actually employ the term *fullness* as a substitution to the term *emptiness*" and see the realization of *sūnyatā* as an occasion of joy (160; 164). As a nondualistic way of knowing, emptiness is actually a vibrant, socially interactive phenomenon: interconnection, not isolation. Instead of freestanding, static entities, everything is said to be interconnected in what the Vietnamese spiritual leader Thich Nhat Hanh calls *interbeing*, such that the so-called self, for example, contains so-called non-self elements. Entities manifest in the moment because of their connection to other entities, in a dependent arising. Hanh's memorable example is of a sheet of paper; mindful inspection will reveal the trees and rain clouds that made the paper possible. Mindfulness means practicing detachment—or not enabling phenomena to become static and freestanding but instead, perceiving their impermanent nature of arising and disappearing. Emptiness is a crucial concept in Buddhist mindfulness practice, because emptiness is what occurs once we detach in the moment. A practitioner of meditation observes an emerging mental formation and returns to open awareness without following that mental formation, a choice that would enable the expansion of that formation into a full-blown storyline. By implication, the perception of emptiness leads to the ending of suffering, which for writers takes remaining flexible to ongoing change in our documents as well as not grasping after writing ability.

The discernment of emptiness is a distinct metacognitive skill that involves seeking out rather than avoiding the nonconceptual—which in turn establishes a mindset sufficiently vacated of preconceptions and evaluation to perceive other possibilities. This takes detachment, not simply as a matter of warding off craving but also the detachment that comes from trying to free ourselves even of arising thought. The intent is to move to a condition of "without-thinking" involving a "non-conceptual or prereflective mode of consciousness" (Kasulis 75). For Bhante Henepola Gunaratana, this "brief flashing mind-moment" is like "peripheral vision as opposed to the hard focus of normal or central vision" (138). Gunaratana adds how nonconceptual awareness counteracts our habit of "focusing on the perception, cognizing the perception, labeling it, and most of all, getting involved in a long string of symbolic thought about it" (138). However, this goal of reaching a nonconceptual state does not mean trying to expunge all thought but, rather, to let thoughts arise without becoming ensnared by them. As David R. Loy explains, "the solution is not to get rid of all concepts, which would amount to a rather unpleasant type of mental retardation" but instead "to be able to move freely from one concept to another, to play with different conceptual systems according to the situation, without becoming fixated on any of them" (228). *Clear consciousness* or *bare attention* refers to the endeavor to see things as they are without adding anything extra—any interpretation or evaluation. In Zen Buddhism, the establishment of this attention happens after a "general house cleaning of your mind" in which "you must take everything out of your room and clean it thoroughly," only returning absolutely essential ideas after they've undergone careful review (Suzuki 110–111). A traditional image compares the mind to a burnished gold surface from which practitioners brush off the dust of arising intrapersonal rhetoric and mental formations (Collins 247). With this image comes an important point: the gold surface is always already present in people—Buddha mind is a shared rather than special attribute and is present with or without mental defilements. In a similar vein, the concept of *luminous mind* in the Indian yogic tradition evokes a glowing, energetic consciousness that makes conceptual activity more apparent by contrast. In a further reduction of set dualisms, luminous mind suggests that consciousness both reveals and evokes mental formations, playing a generative and not a passive role (Thompson 14). Across Buddhist traditions, a range of methods guide practitioners toward experiencing emptiness, varying in time required and self-instruction. Methods include pairing the calming of the mind with an active analysis of the mind for notions of self; exposure to a skilled mentor who "directly introduce[s]" students to a "fully authentic experience of emptiness"; dealing with a progressively more subtle series of challenges to ego the longer one practices the dharma; and the sudden enlightenment of Zen *satori* from *koān* work, from proximity to a gifted teacher, or from a memory from a prior lifetime that leads to the practitioner's awakening (Shonin et al. 168–170).

For a nuanced account of the perception of *sūnyatā*, we can turn to Nyanaponika Thera's *The Heart of Buddhist Meditation*, in which he differentiates between the doctrinal approach (*bare attention*) and everyday application of awareness (*clear comprehension*). As part of the seventh step on the Eightfold Path, bare attention offers a systematic way to handle the mind's tendency to move from its initial receptive state, typically lasting only a few seconds before mental formations intercede, to its second state, in which objects are burdened with labels and editorializing. The role of bare attention is to postpone that second state, "cleansing and preparing the ground carefully for all subsequent mental processes" (Thera 32). The practitioner is usually struck by the freshness of bare attention but will need to work hard to ensure that associations don't accumulate, even after this insight (33). If sustained, bare attention brings three benefits: "it will prove a great and efficient helper in *knowing, shaping,* and *liberating* the mind" (34). With knowing, bare attention fosters metacognition through realization of the otherwise obscure workings of the mind, providing critical thinking capacities such as a "clear definition of subject-matter and terms; unprejudiced receptivity for the instruction that comes out of the things themselves; exclusion, or at least reduction, of the subjective factor in judgement; deferring of judgement until a careful examination of facts has been made" (39). Bare attention assists with both analysis and synthesis—the identification of details and aggregates but also of connectivity, relations, and interactions or the "conditioned and conditioning nature" (35). Additionally, bare attention reduces our tendency toward knee-jerk reactivity by "slowing down the transition from the receptive to the active phase of the perceptual or cognitive process" (35). Finally, bare attention fosters awareness of ongoing impermanence, something people usually only perceive in times of stress or remarkability, and most relevantly, that this impermanence applies to the condition of the self (37–38). In its shaping capacity, bare attention continues to reduce mindless reactivity by "giving to the 'inner brakes' of wisdom, self-control and common sense a chance to operate" (40). It shapes the mind by leashing our urge to interfere in any way—not just recklessly—and by increasing present moment awareness (40). In its liberating function, bare attention provides detachment and insight, including an "inner distance" that might lead to Nirvana by "bestow[ing] upon us the confidence that such temporary stepping *aside* may well become one day a complete stepping *out* of this world of suffering" (43).

While bare attention is a disciplined approach one might associate with a meditation cushion, in contrast, *clear comprehension* as explained by Thera is the application of the Buddhist perception of reality to the daily, routine life of the mind. Thera says that clear comprehension "should gradually become the regulative force of all our activities, bodily, verbal and mental" (45). Clear comprehension is further divided into four kinds: purpose; suitability (choosing the best possible action given circumstance); application of meditation principles to the rest of life to "try to blend it with the work or thought directly required by the day's occupations"; and the continuous comprehension of reality and specifically of a non-abiding self (50). Clear comprehension can be applied to regular daily activities—hence the Buddhist exercises of mindful eating and mindful walking—and for our purposes, it's a green light to apply the doctrine of bare attention to the ubiquitous activities of writing that happen throughout people's lives, in and out of school.

Buddhism offers a procedure to practice bare attention that culminates in immersion in emptiness and no-mind. A series of four meditations or *jhāna* free practitioners from mundane sensory-derived thoughts to freedom from thinking, to equanimity, then to more subtle states in which even the sensations of freedom, mindfulness, or equanimity are gone. In the first *jhāna,* as Ayya Khema explains, a Buddhist "guards the sense doors" and avoids labeling: the practitioner observes the impact of sensory perception, remaining at the level of perceiving without adding anything extra such as categories or slipping off with an alluring storyline. Khema explains that "when the eye has seen the shape, and the mind has said 'man,' or 'woman,'" we stop there. We do not allow the mind to add more. Whatever else it may say will give rise to greed or hate, depending on the situation" (16–17). The first *jhāna* relates to the nonverbal: it strips discursivity of its self-persuasion in a way that also has its uses for prewriting. In the second *jhāna*, the practitioner drops whatever subject had initially tempted categories and turns his or her attention to the joyful tranquility which was evoked by the dropping of those categories. As such, the second *jhāna* speaks to the rapture meditators can achieve, one that "really tranquilizes because it is the antidote for restlessness and worry" (52). In the third *jhāna,* the meditator drops this sensation of rapture to find equanimity, relinquishing any attachment to that meditation joy, a relinquishment which is deepened in the fourth *jhāna* (59). Khema compares the third *jhāna* to lowering the self partially into a well of stillness "just a little, where it is much quieter than on the rim," and the fourth *jhāna* to lowering the self completely into that stillness where concentration is complete and the "ego support system" is dispensed with (65–66). Eventually, a practitioner arrives at the Sphere of Infinite Consciousness, followed by the Sphere of No-Thingness or the cessation of thought (Walshe 162). In the Poṭṭhapāda Sutra, the Buddha explains this process in which "from the moment when a monk has gained this controlled perception, he proceeds from stage to stage till he reaches the limit of perception. When he has reached the limit of perception it occurs to him: 'Mental activity is worse for me, lack of mental activity is better'" (Walshe 162–163).

Emptiness and Mindful Invention

Usually, the writing process is construed as a sequence through which discourse moves from formlessness to ever-increasing form, from multiple possibilities to a stabilized structure likely intended for an audience. This transition is captured in James Britton's discourse categories of expressive and poetic writings, with expressive discourse an unpolished self-actualization and poetic discourse the creation of a structured "verbal object" ("Spectator" 158–159). An accumulation of structural and formal choices—line breaks, patterns in imagery, alliteration, topic and transition sentences, and so forth—give form to the formless, make three-dimensional the one-dimensional, a notion reflected in the idea of a verbal "object." As Britton says, "The change from expression to communication on the poetic side is brought about by an increasing degree of organization—organization in a single complex verbal symbol" (159). Although it is feasible for a text to remain even successfully in one of the earlier phases—journal writing or a disposable freewrite—the majority of first-year writing assignments ask students to advance at least a portion of their work to the later phases. If this urge for advanced-stage, polished writing overwhelms earlier, more exploratory moments in the process, what results is false emphasis on outcome that can lead to problems in composing.

The premise of the Heart Sutra offers first-year composition a nondualistic interplay of writing and nonwriting: writing emerges from nonwriting, and nonwriting emerges from writing. To overprize form over formlessness is to risk writing suffering, since it means that a writer craves a particular product and that he or she is ignoring the present rhetorical situation. In this view, writing perpetually flips to nonwriting and back again—no experience is permanent or static—such that even a highly productive or inspired writing experience will eventually change to nonwriting, and vice versa, even the most seemingly intractable writing problem, if perceived mindfully, will mutate into a stretch of fluency. The impermanence of the situation means there's no reason to dread nonwriting. This dimension of writing has not been addressed in composition theory and pedagogy and perhaps is the cause of significant misunderstanding and stress for individuals trying to write. Formlessness paradoxically serves as content provider: all things arise co-dependently, and emptiness is the forum, so to speak, for vast connectivity and interrelation. The Indian Buddhist philosopher Nāgārjuna, whose *Treatise on the Middle Way* from 150 CE took the philosophy of emptiness to its radical implications, said: "For whom emptiness is possible, everything is possible" (Lopez 30). In mindful composition, form is understood as falling on a spectrum that ranges from the inchoate (nonverbal, sensed ideas without word accompaniments) to the word-by-word tracking of internal discourse not organized for others (freewriting) to genre-specific, highly revised texts. A single letter in Times New Roman font is an instance of form; a well-formed paragraph or use of the Toulmin method is also an instance of form. This approach does not differentiate an occasion of form from another on the basis of perceived usefulness or quality.

It's important to handle the terms we use around verbal emptiness with care: prewriting, preverbal, nonverbal, and not-writing or no-writing. The prefix in "prewriting" and "preverbal" assumes an event outside the scope of the current present rhetorical moment. Specifically, "prewriting" and "preverbal" assume that some sort of written production will happen next in a chain of events, and that the student writer is positioned in a cognitive terrain that abuts text generation. "Pre" connotes that the student writer's current situation of "pre" is ancillary to what comes after the prefix, that "writing" receives more weight and value than "pre." As mentioned earlier, it's pragmatic to talk in terms of prewriting because complete groundlessness could easily overwhelm student writers; however, mentors of mindful composing need to keep in mind the supreme importance of being comfortable with verbal emptiness, of not being in any rush to exit emptiness for words. For the AC/DC spin of form and formlessness to optimally work, both sides need to receive equal valuation. For this reason, I generally prefer "not-writing" and "no writing," because those terms do a better job of honoring emptiness and demonstrating acceptance of activities otherwise disregarded as poor performance. "Nonverbal" performs the service of recognizing the non-discursive factors of a present rhetorical moment. If we resort to calling this set of experiences "prewriting," it should come with the caveat that mindful prewriting isn't the legwork of research, of note-taking, of hunting and gathering concepts, or of discussing a nascent idea with a willing partner. In sum, working with formlessness entails the contemplation of emptiness before language rushes in with the caveat that it would be completely fine if *no language did rush in*, which includes writing experiences that do not come with words, such as embodied sensations. We are not just our words whenever we write.

In the classroom, it's probably most expedient, however, to first understand verbal emptiness through the frame of prewriting and what happens when student writers start assignments. Many student writers have had a hard time managing the blankness that often precedes starting an assignment, and so they may feel uneasy with the idea of reoccurring episodes of not-writing. It's best to tackle how to start writing assignments and demonstrate the benefits of verbal emptiness during that phase before moving on to the emptiness that actually pervades all phases of writing. Prewriting instruction generally seeks to help student writers gain more productive and less stressful experiences with the opening moments of working on documents. Students find starting a piece to be among the most stressful aspects of composing: the incline of this activity is often extremely steep, as though a mental treadmill were set at an impossible angle. A sizeable problem, of course, is the tendency of writers to reach after a chimerical already-written, already-polished introduction or lead. That is, students associate starting with an object that's as of yet non-existent (the first sentence, lead, introduction, etc.) rather than with the temporal and specifically ignore the best example of a beginning—the one immediately at hand in the current writing moment. Ultimately, first-year compositions students should be shown strategies for redefining what it means to start a piece of writing that focus on noticing the newness of the moment and to turn to the moment's verbal emptiness for the switch from formlessness to form.

Prewriting and Emptiness

What is it that writing instructors usually seek for their students' experiences of prewriting? Through writing prompts and invention heuristics, the instructor hopes students will educe tacit material—delineate it, give it form. It seems there are two sets of prewriting desiderata: preparation (that students summon information and organizational options as resources for immediate tasks) and receptivity (that students establish a mindset amenable to new perspectives and experimentation; that students generate a willingness to participate in interior inquiry while understanding the constraints of the assignment or rhetorical situation). Prewriting is among the most astonishing aspects of writing, for the apparent Nothing of the preverbal abuts the apparent Something of writing. One moment, no idea for writing; the next, a workable idea, image, or phrase, and the student writer is off and typing. The cognitive terrain of prewriting recalls a surrealist landscape—polymorphic, with the beginnings of recognizable forms, a mostly empty land in which details are made by chance not ego, by the force of the moment, ones that leave weather-shaped, wind-blown, lone branches, Easter Isle-like profiles, rock outcroppings, tumbleweed, coils, signs at a slant, italicized. No amount of planning and organizing can abrogate the mystery that comes with that moment in which a person moves from no idea to having an idea, from the unknown to the beginning of a sense of direction. This condition is common to writing endeavors great and small, low or high stakes, single paragraph to 300-page draft, irrespective of genre.

Compounding the strangeness of nonwriting in a writing class that occurs with the fragmentary, frequently private, and elliptical work of prewriting, there's a higher quotient of the unknown with prewriting. William Stafford, an American poet whose approach to craft was embraced early on by process proponents, said that writing meant "venturing forth part by part, unpredictable part by unpredictable part" (12). Gesa Kirsch declares "there is an element of unpredictability in all writing" and draws parallels between composing and contemplative practices since both are encounters with the unknown ("From Introspection" W2). In fact, I would argue that writers already encounter a state similar to *sūnyatā* each time they set out to start a new piece, whether or not those writers particularly practice mindful composing. Before we start a piece of writing, we face the unknown and the inarticulate, and we are prepared to face that unknowing in order to ask questions and summon forth material. The transitional moment between nonwriting and writing poses an enigma to first-year students and their instructors alike, for "how can a teacher help a student create form from something that is at present formless, without clear shape?" (Perl and Egendorf 252). Prewriting is sometimes daunting for writing instructors because it means wandering into cognition that's beyond instructors' immediate and assessable control, perhaps bordering on the cognitively invasive. This may be why the field as a whole generally retreated from invention theory after the boom and bust years of invention during the 1970s and 1980s. Comprehensive instruction in mindful invention requires staying attentive to prewriting as a form of emptiness (and thus of interconnection and latent intertextuality), to activities and results which may bear little resemblance to an eventual final text and may not actually involve writing of any kind.

James Britton's focus on "shaping at the point of utterance" concentrates attention on the dividing moment between the preverbal and the verbal and also on what occurs with arising intrapersonal rhetoric as it makes this transition. Without specifically using terms such as "present moment" or "mindfulness" (although "contemplation" and "contemplative" appear several times in the piece), Britton looked at writing as it happens in real time, with emphasis on spontaneity, as a "moment by moment interpretative process by which we make sense" (149). Something is being "shaped" at the moment of divide, and the tools used in that shaping include a person's "store of interpreted experience," real-time sensory experiences, and then the incorporation of both of those sets of material in an endeavor of pattern making (151). The second type of resource used to shape emergent intrapersonal talk is less formed and more of a "felt quality of 'experiencing"—and here Britton draws upon felt sense and "pre-representational experience," a shoe-in for the nonverbal (149–151). According to Britton, this observed content pulls the writer into a state of flow that leads to the generation of further new content, such that "the act of writing becomes itself a contemplative act revealing further coherence and fresh pattern" (151). Interestingly, the emergence of this inner discourse feeds directly into a person's ability to continue writing, stimulating the desire to proceed onward with invention, rather than feeding into a person's tendencies toward correcting and rewriting (148). This view contrasts with the notion that the impulse to edit is automatic and "begins the instant a word has been written" (Mandel, "The Writer" 372). The point of utterance starts off as nonverbal and shifts to verbal, and awareness of this precise occurrence is generative in itself—an example of formlessness yielding form.

In general, theorists have explained prewriting by contrasting it with other moments in the writing process, usually picking a time in which content emerges in a draft, without adequately attending to the nonverbal, with the exception of James Britton and Sondra Perl. In her felt sense heuristic, Perl directs students toward nonverbal emptiness, coaching them to "[b]reathe deeply, repeat the topic to yourself, sense into your body and without writing, see if you can locate where this topic lives in you or what the whole issue evokes in you" (*Felt* 29). On the other hand, Don Murray defined prewriting as "everything that takes place before the first draft" and D. G. Rohman as the activities which occur prior to "the point where the 'writing idea' is ready for the words and page" ("Teach Writing" 2–3;106). As Janice Lauer says of D. G. Rohman and Albert Wlecke, their work with prewriting mainly emphasized the generation of writing rather than the establishment of internal mindsets (78–79). Furthermore, prewriting in their handling reinforced a sense of the discrete self, a form of self-actualization through composing, such that "[w]ithin self-affirmation—the absolute willingness to think one's thoughts, feel one's own feelings—we find the basis for motivation for any writing that pretends to freshness" (*Pre-Writing* 22). Janet Emig similarly juxtaposed prewriting with writing, or "that part of the composing process that extends from the time a writer begins to perceive selectively certain features of his inner and/or outer environment with a view to writing about them—usually at the instigation of a stimulus—to the time when he first puts words or phrases on paper" (39).

Proficiency in verbal emptiness is proficiency in detachment or a willingness to consider alternatives and not prematurely commit to a single approach, to the storyline of one idea, no matter how promising. With mindful composing, student writers explore verbal emptiness to become better creative-critical thinkers and work on dropping preconceptions and premature cognitive commitments, a clearing of the deck that's certainly useful in a range of first-year writing assignments. Metacognitively, this effort can result in an active analysis of their present rhetorical situations in which we remain on the lookout for assumptions and other intrapersonally generated constraints. For instance, students might happily exchange narrow thinking, a static thesis statement, and stunted revision for a consideration of multiple viewpoints, a thesis that moves from working to refined to final, and dynamic and unresolved rewrites. It's advisable that student writers be exposed to verbal emptiness in short bouts through brief, low-stakes exercises as well as in longer sessions—similar to the way in which "[r]epeated episodes of glimpsing emptiness, as well as the steady prolonging of such episodes" is said to guide Buddhists to a "stable realization of emptiness that eventually permeates all of the meditation practitioner's thoughts, words, and actions" (Shonin et al. 166). Operating in verbal emptiness causes a radical acceptance of whatever arises, holding at bay, as much as possible, evaluation and accepting variability, both subtle and substantial, in emerging content and affective responses to the act of writing and learning.

Instructors can work toward these outcomes by attenuating prewriting during high-stakes or formal projects such that the bulk of the composing experience comes from prewriting activities. Encouraging student writers to linger longer in prewriting, these activities shift the focus to accumulating a range of material and perspectives and separate this experience as much as possible from any sorting of the material or choosing between options. In addition, by setting up a nonconceptual ground for their concepts, these bare awareness activities benefit writers by first providing a contrast, making their ideas more sharply outlined and perceptible, and then tilting the mind toward discursivity and invention. Disposable writing is another method in the classroom for preparing students for relinquishing their written products and returning to the position of "having nothing written," a detachment that accepts the consequence of empty-handedness without reservation. Make room in assignments for stretches in which no writing should be produced—like a silent meditation retreat, where the first step is to disengage from talking with other people and the next step is to disengage from talking with oneself—such that students are asked to register intrapersonal rhetoric but not leave a paper trail of it. This bare attention is akin to the progression of awareness that James Moffett wanted to import from formal meditation—the gaining of "some control of the inner stream ranging from merely *watching* it to *focusing* it to *suspending* it altogether" ("Writing" 236). For Moffett, those steps are aligned with how writing "starts in the pre-verbal, with gazing, ends in the post-verbal, with silence, and runs from uncontrolled to controlled mind" ("Writing" 236). Through exercises built around the first two *jhāna*, we can ask students to seek the nonverbal (watching the breath, noticing when self-talk enters, and steering the mind back to paying attention to the sensations of breathing) to arrive at experiences which are temporarily word-free. Encounters with the nonverbal can be arranged through a sequence of descriptive activities—first describing a physical sensation, then describing the cognitive effort to practice *that* description, and finally dropping descriptive language and trying to remain just with the sensation of observing. Attempts at the cessation of inner rhetoric usually strengthen its boomerang return—advantageous for writers. They should gain a sense of the energy of the razed space of prewriting before mental formations jockey to the surface. Traditional seated meditation can also be deployed for the sake of constructing rhetorical situations that have verbal emptiness as their backdrop.

Next is a subsidiary skill of perceiving emptiness—groundlessness—which is the mindful practice of dealing with the continuous alternation between form and formlessness. Groundlessness from a mindfulness perspective means never resting, never settling on one view, because there exists nothing upon which *to rest.* Instead of clinging (often for illusionary security or comfort), a practitioner embraces the ever-changing uncertainty of now and is able to reside in emptiness. As Pema Chödrön puts it, groundlessness means putting aside "our tendency to seek solid ground" in order to embrace ambiguity and uncertainty, leading to "a state of basic intelligence that is open, questioning, and unbiased" (*The Places* 99; 101). Groundlessness also means remaining receptive to possibilities or what Elaine Langer calls "openness to novelty" and "alertness to distinction" (*Mindful Learning* 23). It's dwelling in a state of inquiry through perception of flux and resisting fixed ideas, in a flexibility not unrelated to the one promoted in national writing policies like the "Framework for Success in Postsecondary Writing." Groundlessness is radically contingent and situational—ever-changing and specific to a moment. In this scenario, self and language are historical contingencies rather than timeless, fixed essences, and "we treat *nothing* as a quasi divinity, where we treat *everything*—our language, our conscience, our community—as a product of time and chance" (Rorty 22). What's more, groundlessness is extensive; it persists without beginning or end and applies to all matters, all principles, all concepts, all teachings: even to the very teachings of groundlessness.

For composition purposes, groundlessness is the fullest experience of verbal emptiness, one that requires student writers to relax with groundlessness to explore the non- and preverbal. In becoming comfortable with groundlessness, writing students practice a detached perception of ever-changing context, which in turn encourages the perception of multiple variables and perspectives and suspends conclusions: hallmarks of critical thinking. While contemplating emptiness, things arise, form arises, but that form inevitably returns to emptiness. In practical terms, this means that a student writer tries to not resist constant changes in his or her writing experiences—be okay with impermanence, be fine with having to let go of seemingly attractive arisings of content or self-evaluation as they are replaced by less attractive ones, or by void. In this endeavor, the student writer does not cling to anything, in a radical practice of detachment: everything shifts—content, structure, affective responses to needing to write, embodied sensations pertaining to writing—even writing ability, which is affirmed in one moment but then up for grabs in the next. And it means that a student writer tries to become more comfortable with those moments of void, of not-writing, of the nonverbal. In this second level of endeavor, the student writer does not avoid anything, in a radical practice of attachment—thoughts of failure, skepticism about writing ability, both in general and task specific, content that seems dull or indefensible, and especially stretches of nonwriting.

No-Writing, Not Writing, and the Nonverbal

A tenet in mindfulness practice is the development of awareness of all states, without discrimination or sorting. An encounter with emptiness involves a healthy tolerance of nonwriting: those mind moments normally not categorized as writing-related, including distractions, blanks from the unconscious, impulses and instinctual messages about writing, delays, and affective and embodied experiences. As such, the Buddhist concept of emptiness contributes to composition theory the possibility of a break from potentially intrusive discursivity. Similarly, in his later work, Don Murray adjusted his notion of prewriting to incorporate these mental formations and phenomena not typically associated with written products. Murray's new term, "prevision," entailed "everything that precedes the first draft—receptive experience, such as awareness (conscious and unconscious), observation, remembering; and exploratory experience" (*Essential* 125). After learning how to accept nonverbal elements during writing, the next skill required of mindful writers is their acceptance of the fact that they have been *not-writing*: acceding that this operation in verbal emptiness, one which has led them to see the non-verbal in the writing moment, has in itself resulted in non-production. No new writing has been produced. Technically, by all of the usual measures, the writer has been not writing: a reader will gain absolutely no reading material from the writer in that writing moment. This mindset of not-writing is further divided into types: the acceptance that happens after-the-fact that nothing has been produced from observing verbal emptiness and the intention that is set beforehand to not produce. Frankly, I'd like to teach a whole course in not writing, because it could do a world of good for anxious or perpetually stuck writers. The most important point is that this nonwriting—whether the perception of non-discursive aspects of the moment or the practice of not producing—positions us closer to open awareness and bring us to the formlessness that is necessary for form.

The content of nonwriting encompasses physical sensations, even involuntary and inevitable ones like swallowing, or ordinary sensations like the pain of gripping a pen too hard. That content also involves moments of blank cognition, or a slipping into a Freudian occasion of motivated-not-knowing. Consciousness usually entails a "gappy sequence of moments of awareness," including what experimental psychologists call "attentional blinks," or a failure to perceive a second stimulus appearing rapid-fire after a first (Thompson 50–55). These occurrences don't anticipate the verbal, and during mindful composing we simply rest with what is actually happening. Just as "simply stopping, just allowing the gap, is the first step in the practices of meditation," an important step in developing a practice of mindful composition is acknowledging blanks and gaps (Trungpa *Cutting* 135). Rather than isolate gaps in consciousness, Henry James incorporated them "among the objects of the stream" and categorized gaps based on individuals' cognizance of them (254). Interruptions or time-gaps are occasions when consciousness has completely gone out only to reappear later, and it's possible for people to be aware of these gaps (James 254). On the other hand, what he called quality gaps are "breaks in the *quality,* or content, of the thought, so abrupt that the segment that followed had no connection whatever with the one that went before" (237). These gaps are either "resting-places" or "substantive parts," which appear static and unchanging, or "transitive parts" and "thoughts of relations" that function as bridges between the static gaps. James compared those bridge moments to highly relational grammatical elements such as prepositions, conjunctions, and adverbs, recalling that web work of connections in Buddhist emptiness, and he proposed that individuals should develop awareness of those moments (249–250).

The benefits of becoming aware of the involuntary are probably most obvious with mindful breathing, an involuntary act that is consciously observed and given the appurtenances of the voluntary, yielding insight and connections. It's possible to evolve consciousness of unconsciousness, evinced in lucid dreaming, hypnagogia, and yoga nidrā (sleep yoga), so the skilled practitioner is able to meditate while dreaming, as Evan Thompson has pointed out (110). In fact, a role is allotted to the unconscious as a nonconceptual mode in both Buddhist emptiness and preverbal emptiness. The unconscious joins formlessness and non-abiding as the three central tenets of Buddhism (Suzuki 189). In Zen Buddhism, stretches of the unconscious indicate an ultimate condition of detachment in which the particles of experience do not adhere to the surface of the mind, "to be always detached from objective conditions in one's consciousness, not to let one's mind be roused by coming in contact with objective conditions" (Hui-neng qtd. in Suzuki 189). This is an interesting addition to the perspective on arising mental formation; typically, mindlessness is diagnosed as the result of becoming ensnared in storylines built by inner discursivity. With the Zen approach to the unconscious, however, what is praised is the de facto way in which the unconscious by its very make-up prevents detachment: no discernible intrapersonal rhetoric is generated, and the manipulations of the unconscious pass unregistered. The mental formations are contained in the individual without manifesting—"to have thoughts and yet not have them"–and that's why the emptiness of the *prajñāpāramitā* is equated with unconsciousness in Zen (Suzuki 189–192). Here the unconscious also leads to a state of no-mind and liberation from the trappings of ego. In terms of writing, D. G. Harding posited a set of mental formations that occur beyond conscious control and even precede intrapersonal rhetoric in a "hinterland of thought." This type of awareness involves "emergent impulses and processes that ... may exist in modes far different from words or imagery" (Harding 177). To be unconscious and preverbal is to be empty—polished gold without dust.

A mindful way to manage the formlessness side of that interplay between form and formlessness is to conceive of formlessness as tacit knowledge. In other words, how might student writers grasp the idea of a vast interconnected intertextuality, of a co-existence of words and wordlessness? How might they become less wary of blank pages and blank screens and more trusting that turning toward rather than away from verbal emptiness will yield content for their pieces? With tacit knowledge, material exists outside of our conventional consciousness but is nevertheless already around. It becomes part of an interconnected limitlessness. We can tacitly know something without it standing within our explicit analytic reach; we may know something that is nonconceptual, that somehow retains traits of the unknown in that we lack words or a voice to discuss it. Michael Polanyi says, "we can know more than we can tell" (4). For Polanyi, the kind of knowledge which is explicit is only discernible because of the tacit, and on the other side, once we arrive at meaning, "all meaning tends to be displaced away from ourselves" (10–13). This idea is similar to Sondra Perl and Arthur Egendorf's notion that the nonconceptual means sensing the presence of something that wants to be written but not yet having the words for that material, which they call "knowing but not knowing" (251). Perl and Egendorf observe, "Even single words like 'ineffable,' 'inchoate,' 'implicit,' can point to the presence of 'something' we experience that extends beyond words" (254). Likewise, James Moffett located a tacit dimension in composing when he calls inner speech "an uncertain level or stage of consciousness where material may not be so much verbalized as verbalizable, that is, at least potentially available to consciousness if some stimulus directs attention there" ("Writing" 231–232). Early discussions of prewriting connected it to tacit knowledge; D. G. Rohman claimed that "writers set out in apparent ignorance of what they are groping for; yet they recognize it when they find it. In a sense they knew all along" (107). By investigating formlessness as a type of tacit knowledge, a mindful theory of composing joins the branch of invention theory in Composition Studies that already looks at whether "inventional practices are non-discursive acts or are symbolic" and includes tacit knowledge along with explicit knowledge (Lauer 2). It's formlessness while recognizing the imminence of form; it's wordlessness that contains a word.

Nonwriting has a special function in writing, and this function should diffuse the usual self-recrimination around not producing words. Don Murray proposed five natural delays experienced by writers as waiting for voice, insight, organization, purpose, and information. With each reason for delay, a writer appreciates the more inchoate impulses and needs experienced by writers before they are able to start a draft, including matters like possibility, hints, guesses, exigence that fulfills the self's needs, and an intrapersonal voice ("The Essential"). As he says, "The writer has to accept the writer's own ridiculousness of working by not working" and "accepting the doing nothing that is essential for writing" is key to the development of a piece (226). Nonwriting isn't just a matter of relying on the unconscious or giving up control; it's really a matter of noninterference, of not meddling with the details of a present moment. Instead of adjusting a situation to our comfort, by nonwriting we accept the present and seek "to open and relax without adding anything extra, without conceptualizing, but to keep returning to the mind just as it is, clear, lucid, and fresh" (Chödrön *When* 20). Contemplating the not-writing side of prewriting shrinks our usual tendency to interfere by "doing something"—compel a draft, get started, freewrite, produce words, any-words-as-long-as-we're-writing. This notion may seem counterintuitive to established views in composition that encourage low-stakes, informal, exploratory writing (like freewriting)—a "just do it" mentality that tries to help students over the hump of starting by lowering standards. Undoubtedly, the endeavor to lessen the stakes does much to support student writing; it nevertheless carries an implied valuation of product (get *some* sort of wording down), perpetuating mindlessness by avoiding an element of the present writing moment (the nonverbal). This connects to the Zen notion of not adding anything extra to the moment, including false evaluation, which, as concerns the act of writing, means something as basic as valuing writing activity as valuable and praiseworthy and not writing as worrisome and even punishable. Writers who are better able to balance the constant fluctuation between form and formlessness and appreciate rather than reject the formlessness side of the interplay are less likely to suffer from a host of typical writing problems, less attached to outcome or particular results, comfortable knowing that any writing situation will eventually, even in the immediate moment, shift. Craving written products and final outcomes (finishing a demanding task, receiving a particular grade) inflates the value of form and risks writing suffering.

Clutching and releasing is a familiar strategy in both contemplative practices and contemplative composing pedagogy for contacting emptiness. Clutching—whether of muscles, emotions, or concepts—draws attention to habits of retention and resistance, and the contrast shortly afterward with purposeful release builds a sense of openness. A writer, for instance, can fix a strong grip around a working thesis statement or his or her vexation about a deadline and then purposefully and fully release all thought pertaining to that mental formation. In Vipassanā meditation, this clutch-release pattern happens when practitioners are instructed to transition from a focus on the breath to a more open awareness (Thompson 52). In *Meditation as Contemplative Inquiry*, Arthur Zajonc describes "cognitive breathing," which entails shaping the mind's effort into a two-part formation, like inhalation and exhalation, except that the first is a focused attention followed by an open attention. The practitioner hones her attention entirely on a certain content and then completely releases her attention on that content. What results is an "open, non-focal awareness" that resembles formal meditation in that "[w]e are entirely present. And interior psychic space has been intently prepared, and we remain in that space. We wait, not expecting, not hoping, but present to welcome whatever experience emerges into the space we have prepared" (39). Another example is how, approximately midway through Sondra Perl's procedure for composing with felt sense, after students have written on topics such as what's on their mind, she tells them, "I'm going to interrupt you and ask you to set aside everything you've just written … to set aside the bits and pieces" in order to connect to a greater whole through felt sense (29). Basically, students focus on their intrapersonal rhetoric (the material generated by her prompts) and then release that focus to discover an idea residing in the wake of the nonverbal. Perl and Egendorf explained the preconceptual in a grasp-release pattern similar to cognitive breathing: "If you 'drop' your attention from the particular words or images that are occurring to you right now, you may note a global, perhaps ineffable, yet nonetheless unmistakable sense or mood—just under the words or images—that encompasses *all* of how you are at this instant" (255–256). Paradoxically, binary activities (like clutching and releasing, talking and sitting silent) undo some hindrances typical of binaries.

Emptiness of Writing Self

Part of noticing verbal emptiness is registering how inner discursivity tries to establish a sense of self, that illusion of an autonomous self, which in the Buddhist perspective leads to problems. As Daniel Lopez explains it, emptiness isn't "the negation of existence but rather is the absence of a particular kind of existence, an existence that is independent of any other factors. If nothing exists independently, then everything exists dependently and lacks, or in other words is empty of, independent existence … a middle way between the extremes of existence and nonexistence" (29–30). With an illusion of a fixed self, dynamic interconnection is reduced to what Chögyam Trungpa called a "frozen house, made of frozen colors and energies" once the ego invades vast openness (*Cutting* 129). In the previous chapter, we discussed no-self in relation to intrapersonal rhetoric to counter the assumption that a writer's voice is somehow indicative of a permanent expressive self. With *sūnyatā*, we gain a better idea of the "location" of that no-self as it retreats from the center stage of ego to where it joins hands with all other performers, entities, and phenomena.

Buddhism fosters meta-awareness, such that individuals can be cognizant of the frequent occasions in which ego is ensconced. As Evan Thompson describes it, open awareness lets individuals "see how habits of identifying with sensations, thoughts, emotions, and memories—in other words, with specific contents of awareness—create the sense of self" (52). Referencing Thomas Merton, Arthur Zajonc explains that creating a clear mind not only leads to receptivity and acceptance for emerging events but also helps the practitioner reach egolessness: "In turning my attention towards this silent self, I sense the intimations of a Self that is no-self" (38). Similarly, George Kalamaras, speaking of Hindu and Buddhist mysticism, describes a "generative paradox" that comes with the clear mindedness of meditation in which "an 'emptying' of the consciousness of a limited self" is "'filled' with the perception of the dynamic interaction of all particles of creation," resulting in a mutually supportive dynamic between the social and the private (*Reclaiming* 21). The concept of emptiness recognizes "the clinging to the notion of self—as one of the most severe illnesses suffered by humankind" and that the role of emptiness lies "in performing its soteriological function—which is to help sentient beings sever their attachment to the self" (Low 133). Perception of emptiness is salvic, as Lopez says, because it allows practitioners to relinquish the notion of a fixed self and become "stream-enterers," a kind of intermediary-level orange belt disciple who has dissolved the self (48–50). It's a status change evident in the Heart Sutra's account of practitioners transcending consciousness—"he has passed beyond discrimination and knowledge, beyond the reach of change or fear, he is already enjoying Nirvana" (Goddard 86). Buddhist emptiness refutes notions of the essential, fixed, or enduring and replaces such ideas with the socially constructed and situated concept of dependent origination. This philosophy of radical emptiness reached its fullest implications around 150 CE when Nāgārjuna, an Indian Buddhist philosopher, proposed a happy medium between "extremes of existence and nonexistence" in the *Treatise on the Middle Way* (Lopez 29). Nāgārjuna defined emptiness as a vast network of connections; the self is empty because it is "simply the locus of all these causal interactions" and "everything disappears under ultimate analysis and has only relational existence" (Priest 470). Maintaining the groundlessness of the Heart Sutra, later Buddhist thinkers include the teachings in that relational, interdependent matrix (Priest 469). Since formlessness links all entities, it becomes a collective and highly networked space where our individual egos dissolve.

A seemingly solid entity, including the self, dissolves into a mist of aggregates which are in turn associated with other entities beyond the initial entity. Imagine a tangerine as it becomes a Dali-esque collage of sensory information and organic and commercial histories in a composition that is constantly in transition. In *How to Meditate*, Kathleen McDonald explains that emptiness is "emptiness of inherent existence" and that we mistakenly "see things as fully, solidly existing in and of themselves, from their own side, having their own nature, quite independent of causes, conditions, parts, or our own mind experiencing them" (53).

Essentially, acknowledging emptiness involves acknowledging our subjectivities and their shaping influence on perception and understanding. McDonald offers this example:

> Whatever exists is necessarily dependent upon causes and conditions, or parts and names, for its existence. A car, for example, is a collection of pieces of steel, glass, plastic, rubber, and an engine, etc., put together by people in a factory. We give the name 'car' to this collection, but if we look for a real, concrete, independently existing car, it cannot be found.
>
> *(56–67)*

Her example is a variant of the classical Buddhist parable of the chariot in which a king named Milinda is quizzed on his vehicle. Asked whether he arrived on a pole, wheels, or an axle, Milinda is coached into realizing that no such entity as a "chariot" exists, only the composite of its dependent qualities. Buddhist disciples were asked to contemplate how nothing is freestanding or uninvolved in multiple contexts. Nothing is solid. Instead, all entities are a collection of aggregate factors, each factor connected or related to something beyond itself.

The only path to reaching this shared Buddha Nature, this non-differentiated "you" and "I," is by diffusing binary thinking. Since formlessness joins all entities, it is a collective and highly networked space where our individual egos dissolve, recalling Nāgārjuna's radical relational existence. In his discourse on emptiness, Nāgārjuna used a syllogistic verse to cover a range of topics including motion, fire and fuel, bondage, agent and action, limits, self, nirvana, errors, combination, enlightened beings, time, and the Four Noble Truths. Two syllogistic poems are "Action depends upon the agent. / The agent itself depends upon action" and "Without detachment from a vision, there is no seer. / Nor is there a seer detached from it. / If there is no seer, / How can there be seeing or the seen?" (Priest 478). In Composition Studies, this view of self is evident in classroom approaches that deemphasize a static sense of a discrete student in pursuit of teacher acceptance of their writing and a satisfactory final course grade or of writing acts as autonomous self-expression. For example, in her praxis of connected knowing, Christy Wenger illustrates how embodied imagination helps students "experience the self in relational webs" and "attend to others' [thoughts and feelings] whether real or anticipated" (*Yoga* 25). In developing antiracist writing assessments, Asao B. Inoue applies Buddhist interconnectedness to help students and teachers problematize white racial *habitus* as interdependent with less dominant discourses in their classrooms, underscoring the need to examine "language and judgment through their differences" (*Antiracist* 93).

In a broad sense, when students are able to drop fixation on a single social dynamic in their writing experiences (for instance, an anticipated intimidating critic), they can instead appreciate the myriad of other ways, great and small, they're connected to others through their writing and education. The drama fomented as the discrete ego goes into battle is balanced with reflection on how writing is sustained by a myriad of other interactions and relationships. These might include a moment of kindness from a fellow student, the extra twenty minutes a teacher gives help outside of office hours, the gift of language itself, which the student shares with all other users of the language, and the writing materials and technologies which workers built in factories. Conversely, a sign of a false binary between self and social is evident in composing theories whenever invention is described as either the discovery or the construction of material. Does attention to preverbal emptiness cause it to be generative, or do writers dip into their intrapersonal stock and locate preexistent material? Does new content emerge from this attention, or does mindful composing simply encourage people's natural habit of generating inner language?

Mindful composing theory addresses an ongoing epistemic question, one resulting from a dualism of self and other that has long steered the general direction of invention theory. Composition-Rhetoric scholars "continue to debate whether rhetorical invention helps writers to construct new knowledge or only to find arguments or material to support and convey judgments reached elsewhere (e.g., through philosophy or science)" (Lauer 3). Tropes of invention, discovery, and creation—ubiquitous in writing theory—are not interchangeable but, rather, show "three quite different orientations in understanding discursive production" (Young and Liu xiii). Discovery and creation establish a false binary in which discovery posits a "preexistent, objective determining rhetorical order" whereas creativity suggests "a generative subjectivity as the decisive factor in initiating and sustaining the writing process," and invention dissolves this binary by designating a "uniquely rhetorical perspective on composing that subsumes both objectivistic and subjectivistic conceptions" (Young and Liu xiii). A cocktail of "made" and "discovered" is evident in Rohman and Wlecke's division of prewriting into four activities: "effective concept formation"; "attention to the particularity of events"; "attention to the personal sense of what is real"; and "transference of an already known structure of events to the events of the subject," a list suggesting self-actualization plus agency plus preexistent patterns (7). Mainstream writing education opts for the "discovery" side of the binary, such that writing is viewed in "relatively simple terms as a technology for communication and a straightforward, rule-governed process of encoding a more-or-less stable meaning in a text" with the result that "in teaching writing in conventional ways, we are teaching a dualistic way of being in the world" (Yagelski 12). Yet, from a mindfulness perspective, questions of whether writing is a matter of creativity or repurposing of found material become rather moot, given the absence of a discrete self and the fundamental interconnectivity of entities. It's the cyclical movement between binaries that's generative; the verbal fills a present rhetorical moment until nonverbal elements inside the verbal become dominant, and vice versa. A striking example of binary-busting in Mahāyāna Buddhism is the equivalency established between nirvana and samsāra (endless cycle of birth, suffering, death), which are considered one and the same, such that nirvana contains samsāra and samsāra contains nirvana.

I can't imagine how the stakes could be higher in a spiritual system than in discussions of the afterworld, yet in Buddhism even those weighty topics are given equivalency—surely smaller dualisms concerning writing can be dealt with.

Catching a glimpse of the nonverbal and then keeping it at the forefront of our thoughts stops the landslide through the five *skandhas* leading to the development of a falsely discrete writing ego. In *Cutting Through Spiritual Materialism*, Chögyam Trungpa documents the installation of ego in emptiness through the five *skandhas.* Our original or primary state is one of openness: "Fundamentally there is just open space, the *basic ground*, what we really are. Our most fundamental state of mind, before the creation of ego, is such that there is basic openness, basic freedom, a spacious quality; and we have now and have always had this openness" (122). In this original nature, along with the Edenic, there's the notion of an early condition of vastness, which Trungpa calls a "primordial intelligence connected with the space and openness" (123). Paradoxically, it's our awareness of this vastness that excites activity which then limits that vastness. Trungpa uses the metaphor of a dancer in an immense dance hall: the dimensions of the space remain, but our movements draw attention to ourselves and to ego: "Because it is spacious, it brings inspiration to dance about; but our dance became a bit too active, we began to spin more than was necessary to express the space. At this point we became *self*-conscious, conscious that 'I' am dancing in that space" (123).

In the First Skandha, individuals solidify that openness as a way to ignore or deal with the openness that only became overwhelming after we mistakenly propped up an ego. This problem happens if the person notices a separate self and then believes that the ego situation has always been as such. The ability to perceive oneself fortifies the false notion of the self as a discrete other. In the Second Skandha, the effort to solidify space extends from the self to items in the surrounding space; through use of the five senses and through evaluation, these items become another. In the Third Skandha, individuals become fascinated by these separate entities, which they have in actuality generated through their own perceptions. In the Fourth Skandha, individuals name and categorize these entities as a "defense to protect one's ignorance and guarantee one's security" (127). In the Fifth Skandha, consciousness is established as the products of the other four skandhas unite, and thoughts and emotions are generated. In first-year composition, student writers can work on developing mindful meta-awareness by learning to notice how their intrapersonal rhetoric persuades them into certain positions, anticipations, and moods and ensconces writing ego.

In a group of unusual articles published in composition journals in the late 1970s and early 1980s with titles like "The Writer Writing Is Not at Home" and "Losing One's Mind: Learning to Write and Edit," Barrett J. Mandel broached this notion of a writing no-self. "No one is home" when writing symbolizes an evacuation of consciousness that happens because, Mandel claimed, writing occurs through the body without contributions from the mind, such that words "appear on the page through the massive coordination of a tremendous number of motor processes, including the contracting and dilating of muscles in the fingers, hand, arm, neck, shoulder, back, and eyes" ("Losing" 365). The writer metonymically disappears into her writing implement: "More accurately, I *become* my pen; my entire organism becomes an extension of this writing implement" (365). Adapting the ideas of Maurice Merleau-Ponty and Julian Jaynes, Mandel suggested that consciousness is a projection or analog rather than the direct source of thought. His description is reminiscent of Buddhist notions of the mind-made and of the endless emergence and disappearance of mental formations on the screen of the moment: "Consciousness is the illusion—the make-believe—we call reality projected upon the screen we look at. Upon this screen in any given moment is projected everything we call real: our thoughts, physical presence, objects of contemplation and perception—everything" ("The Writer" 371–372). This consciousness is an evasive phenomenon, since it suggests its own passivity and non-involvement, similarly to internal rhetoric, which pretends not to be there but actually enacts sweeping campaigns on our actions, behaviors, and moods. As Mandel says, "[c]onsciousness does not create or generate; it merely keeps the record," but in projecting these formations on the screen of the mind, it also omits the fact that it is doing the projecting (374). Similarly, in more recent research, Rachel Forrester describes the strange independence from self and from conscious control that occurs during writing and the paradox of how the work of writing ultimately calls for a sudden end of trying, despite all sorts of planning. In its propensity to remain an invisible influence, consciousness tends to perpetuate mindlessness. We can't assume the reliability of consciousness, because it also evades, pretending it plays no role in constantly arising mental sensations—a hallmark of mindlessness in learning.

We need nonwriting for the purposes of writing. We need to teach students to be less wary of the blanks (pages or moments), because a blank contains everything. We can perceive verbal emptiness in every breath and during every present writing moment. Far from worrisome or isolating, observed emptiness is the moment in which writers are connected to everyone and everything, and all sorts of possible content for their writing is within reach. Emptiness is writing at 100% rather than in a zero or void. Whenever we write, we make contact with countless moments in which we are *not* writing: any piece of writing only emerges because of emptiness.

Interchapter 3

Write Your Own Sand Mandala

Write your own sand mandala—writing that is blown away. Reach beautiful insights, find colorful patterns of ideas, structural strategies, realize new points and segue, create whole stretches of writing that turn at margins and fill another line—and then erase. Write as you would normally write or write as you would not normally write, but at the end of the writing session, delete. Write with an audience in mind or write with no audience in mind, and at the end of the second day, shred. In teams or as a single author, give yourself a focus, genre, approach, or do not give yourself a focus, genre, or approach and instead freewrite, and at the end, crumble, shred, press delete. Write the next step in a draft on a particular project or begin something new. The content, stage in process, and genre do not matter—but in the end, delete. After deleting, reflect for a few minutes on the new present rhetorical moment, a space razed of this written product. A mandala of writing is like a Buddhist mandala that "comprises a cycle of allowing the practitioner to build up hopes and concepts, only to have them abruptly dismantled" (Shonin et al. 169). As instruction in groundlessness, the mandala of writing helps first-year students detach from already-produced content (including their built-up hopes and future considerations for a piece) and rest more comfortably in not "having" anything written. By not overvaluing writing ability, individuals grow less fearful of writing inability; not caught up in that fear, they are better able to perceive the present rhetorical moment and its affordances, including how formlessness changes over to form. The mandala is an emblematic praxis of verbal emptiness and an occasion of instruction in groundlessness, one of many in a mindful composition course.

In the classroom, practice with verbal emptiness directs us to two learning outcomes: first, to learn to appreciate and not avoid blank moments of seeming non-production, and second, to learn to examine those blanks to locate possible content and be an observer of the changeover from formlessness to form. (The stipulation is that writers also learn to notice and accept when this situation reverses—when form returns back to formlessness.) While acknowledging external pressures of assignments, deadlines, and grades, we want student writers to rest more comfortably with writing uncertainty, knowing that the present rhetorical moment is expansive, accordion-like, and if observed, allows calm observation. It is of critical importance that instructors help student writers find ways to reduce the constant expectation of themselves that they produce, and as such, we need to remain careful as writing instructors of the pro-production messages sent by well-meaning heuristics of abundance such as freewriting. It's not just a matter of lowering writing standards and therefore of diminishing evaluative tendencies in intrapersonal rhetoric. Instead, working with verbal emptiness means not expecting any production in the first place—altogether blocking any possibility of judgment.

The assignments need to send the message that prewriting and nonwriting are acceptable activities and at any stage, from starting to editing. Otherwise, students might naturally associate these experiences with negative consequences such as a failing grade or a drop in their teacher's opinion of them. With verbal emptiness, instructors develop tasks that let student writers show their encounters with prewriting and nonwriting, to legitimize emptiness as a classroom activity. Verbal emptiness assignments resemble more traditional assignments in terms of accountability, feedback, and evaluation. In addition, classroom time needs to be allotted to discussing students' affective responses to these likely unusual writing circumstances (when they are encouraged to linger in prewriting or embrace nonwriting): for instance, some students' unease as they face the uncertainty of this new educational experience.

Disposable writing of any sort—whether intricate like a mandala or as informal as a freewrite that's immediately deleted—is a way to practice detachment and build willingness to be nonverbal while writing. One student professed his fondness for disposable writing, saying that disposable writing "seriously works wonders for me. I love being able to write anything and then just have it disappear" (Marcus). Clinging causes writing suffering: it's the ego attempting to control the uncontrollable by limiting its own choices, opting for the perceived security of a single thought or vantage point. Student writers frequently seek a quickly arriving definitive as the basis of their writing because openness to a wider range of possibilities can seem chaotic and threatening to their academic success. The verbal emptiness that results from disposable activities during mindful invention differs from the deletion that frequently happens when writers conflate creation and correcting by hitting the backspace as they compose. While both result in the absence of written production, the latter is a defensive move caused by the false perception of an imaginary audience in a rhetorical moment during which emptiness isn't a desirable state but, rather, signs of an anxious mind. Early on, first-year writing students find the disposable method most useful and least intimidating in conjunction with freewriting, momentwriting, or the earliest stages of invention. One student limited the disposable activity to a paragraph to manage the openness of prewriting: "I wrote a disposable paragraph to see what is really in my mind, without editing as I wrote" (Jenny). Another student purposefully selected materials (Post-It notes) in conjunction with disposable writing: "For the second prewriting exercise, I chose disposable writing, and so I took out my Post-Its and started by watching my breathing, writing little stories on these notes. I wasn't thinking too hard because I knew I would throw them out. I actually liked what I said on some of the notes, so I wrote down some of the content and tossed the rest" (Allison). Advanced students can practice with more polished phases and with genre of increasing distinction, such as a writing mandala, or extend disposable writing sessions from an activity to a Save Nothing Day to perhaps even a Save Nothing Week or assignment unit (and submit a process essay instead about the experience). In yet another paradox of mindful writing, disposing of writing is probably one of the best ways to help severely blocked student writers. If a piece of writing is declared ahead of time as lacking in value (i.e., it'll be immediately trashed), the student is less inclined to engage in intrapersonal debates about the quality of the work, which in turn frees the student to perceive arising content and create writing they'd like to keep. On one occasion, I have watched as a student suffering from a disabling block that made it nearly impossible for her to write in a classroom setting begin rapidly writing a few feet away from a prominently displayed document shredder—and surreptitiously began tucking pages in a notebook, thinking I wasn't noticing. Of course, I let her keep her "disposable" writing after discussing what allowed that writing scenario to yield not only writing but writing that she wanted to retain.

Another method to engage with the pre- and nonverbal dimension is what I call *momentwriting*, an alternative to freewriting that maintains present awareness for writing without the mandate that words be produced. With freewriting, the principle is that a student tries not to stop writing. With momentwriting, the only rule is that a student tries not to stop observing the moment, including the breath. Instead of asking for non-stop writing that pushes over and past blank moments, momentwriting sees blanks (or physical sensations) as worth recording in writing: the cry of a blue jay, the aftertaste of coffee, a sudden wave of wordless energy, a wordless image, a passage in which mainly the breath is noticed. Momentwriting allows people to track impulses and sensory experiences and to honor them as part of their writing experience. For instance, if while momentwriting I feel a strong impulse to consult a certain notebook on a shelf, I let myself stop typing and retrieve the notebook while maintaining mindful breathing. I've found from years of writing practice that there is usually a sound reason for the impulse (a forgotten source or a relevant phrase), one relevant to the writing task at hand. On a few occasions of momentwriting, I've even located an exact thought in an old notebook I haven't opened for a decade. An impulse could be as seemingly trivial as suddenly wanting to change pen colors (these impulses often parallel important shifts in the content of a draft). Similarly to freewriting, momentwriting lowers the standards, because as long as student writers continue to watch their breathing during activity intentionally designated for writing, no matter if they produce sentences, a fragment, or nothing, no matter if what they've noticed is as nonverbal as the sensation of the ballpoint pen, it's a successful momentwrite.

By now, it should be clear that momentwriting incorporates rather than avoids the nonverbal or silences. Like freewriting, momentwriting engages students' intrapersonal rhetoric, but unlike freewriting, it does not privilege discursivity over other aspects of the present moment such as blanks from the unconscious, nonverbal sensory experiences, emotional waves. Unlike freewriting, momentwriting doesn't goad the hands to keep up with a certain handwriting or typing rate in a "Just Do It" fashion but, rather, lets the pace happen more naturally with intrapersonal talk and emerging mental formations and doesn't exclude other present rhetorical factors. An undeniable benefit of freewriting is how it forces writers to just keep moving and not stop to fix; its forced march also tries to keep up with the natural quick flow of usual internal thought without sorting. However, in keeping up this fast pace, freewriting usually omits nonconceptual real-time experiences and overly values the verbal.

Like freewriting, momentwriting navigates from the left to the right margin, preserving the intrapersonal rhetoric as a text, giving the appearance of permanence of a piece of writing, however fragmentary. Students use visual elements to record the nonconceptual parts of a momentwrite, such as blanks made from tabs or using brackets, parenthesis, or italics for material typically omitted from a freewrite, such as a fixation on shoulder tension for a few seconds. For example, in this excerpt of a momentwrite, I've used brackets to indicate a lull where I've left the writing and backslash to indicate moments in which I was aware of a physical sensation related to the posture and effort of typing without putting the sensation into words (on other occasions, words did arise for a physical sensation):

> Reason why my impulse is to change pens mid-stream during a writing session—moving from Bic ballpoint to magic marker to dollar store mechanical pencil, from black to blue to pink to green—is to reflect (and capture) *demarcate* changes in time [] that it is a new moment, that the phrase or idea is on a distinct flow in that intrapersonal babble passage. This tea tastes fruity. An attempt to not be unified not hold writing together at this early stage of invention. ///// To do so suggests undue precautions taken for considerations given to an unknown and future audience. And that means departing from moment. ///// For a long time, I have wondered though without doubting or challenging or correcting it why I have this inclination. In a passage like this one where I don't change pens—I'm pounding the keys too hard—state of flow "inspiration," glued together with more continuation through voice, sense of non-stop moment, this tempo of getting it all down, that shoulder is hurting. Sometimes change in pen though a form of evaluation (did I forget about that email?) to highlight potential excellence—so to evaluate, remind myself of what to (later) pursue.

A message reinforced by momentwriting is that the writer's entire internal experience constitutes acceptable composing work: freewriting went to considerable lengths to send this message of acceptance, but momentwriting constitutes arguably more radical writing self-acceptance. Certain mental formations, such as recalling a work email or wanting to use a different pen, are not labeled distractions, and as a result, the writing experience is that much more evaluation-free. Similarly to freewriting, momentwriting can be focused or unfocused, shared or private. While it's likely that momentwriting will generate a text of some sort, however elliptical and disorganized for a reader, with at least a modicum of usefulness, it's possible that observing the breathing and momentwriting will yield nothing worthy of being kept or possibly no language whatsoever. Just as freewriting keeps success at a low bar (whether one kept writing, regardless of quality), momentwriting puts the parameter of success at whether or not one kept watching the breath with the intention to write. To borrow a laboratory scientist's slogan, "Another Day, Another Fail": the attempt is what matters.

Practicing cognitive clutching and releasing in order to develop bare awareness can be accomplished through "Corpse Pose for Revision" (also called "Relaxation Pose for Revision"), a *jhāna* modeled after the beloved *savasana* pose at the end of yoga classes. Preconceptions about one's writing can obstruct progress and revision. When students can't effectively continue work on a piece because they are essentially preoccupied with worry or hopes for a particular outcome for the writing, they can restore themselves to a more open, inventive position. "The Corpse Pose" reduces preconceptions around a draft by asking students to focus on an area in the draft and then completely release that part from their consciousness, resulting in bare awareness. This method returns students to a mindset of formlessness and invites them to face verbal emptiness even once (especially once) they have a collection of written material.

Students begin by clearing their writing area of any signs of the writing project, including pens, pencils, Post-Its, notebooks, and relics of feedback or grades. Next, they divide the draft into five to seven parts, each not exceeding 250–500 words or a piece that can be read with ease within a minute or by its paragraphs. Dividing the work may entail excerpting from a much longer document, and if that's the case, the student should select sections which seem troublesome. Each segment is placed on a separate screen or printed out on a separate page: moving in reverse order, the chunk closest to the end of the document (the feet) is moved to the first screen or sheet of paper, followed by a subsequent passage on the next screen, until the very last screen or page of paper holds the opening (the head) of this document. Watching their inhalation and exhalation, students turn their attention to the "feet" of the document and only the feet. They should put all of their attention on this section and reread it, scanning it for any tension that arises. *Where are you frustrated, irritated, worried, or any other emotion?* Students should not ward off or suppress these emotions but, rather, simply observe them with a detached mind. They scan for images, associations, and new ideas that arise from mindfully watching the feet of the document. After a minute, they release this part of the document: release "the feet," letting it sink back down onto "the floor" (sheet of paper). If working on a computer, students should close the screen holding that piece of the document. They let go of everything—all mental formations and sensations—concerning that section.

Next, watching their inhalation and exhalation, they direct their attention to the "calves and thighs" of the document and place all of their attention only on this section, rereading it. They follow the same steps as for the "feet," ending by releasing all mental formations pertaining to that segment of the draft. The same procedure is used on the "pelvic area" and "belly" of the document, followed by the "torso" or "chest," the "arms" and "hands," "shoulders and neck," and the "face." With the "face" or introduction of the document, students try to observe even the most minute mental-musculature tension pertaining to this section. Because the introduction is the most noticeable part, the "face" often contains complicated stresses, built up over time. Lastly, students move to the "crown" of the document, the space above the first section, perhaps where a title stands or might reside one day. By now, students are probably more relaxed, their affective constraints altered, and are more cognizant of their assumptions about the draft. Describing her experience with "The Corpse Pose," Amanda used the method after the more additive Montaigne method in an intriguing pairing of additive activity with relinquishing: "I cleared my desk, and by clearing my desk, I was able to also clear my mind," releasing each part of her draft. Students should linger a minute or more in this state and, if possible, have another student ask a question about their draft or writing experience. In this relaxed state, so close to the floor, so to speak, so close to the unconscious, the student may find insights and ideas not possible with a strained, tight mind.

More formal writing projects that significantly expand the portion of the composing process allotted to prewriting offer students increased practice with verbal emptiness. Instead of confining prewriting to an initial discussion or a day's worth of prompts, this approach makes prewriting the bulk of the writing experience. For example, I assign a three-week-long self-directed project in which students decide nearly all aspects: genre; length; topic; and for the most part, who has the chance to read the piece (and when they read it). Instead of submitting regular drafts, students submit a log entry before each class meeting in which they detail their daily writing experience with the project, drawing their attention to the present writing moment as they consider factors such as audience proximity, intrapersonal rhetoric, preconceptions, and affective responses to the task. Sometime during the three weeks, students need to pick from a range of prewriting options (including freewriting, momentwriting, yoga for hands, and disposable writing); invention strategies (including freewriting, momentwriting, Peter Elbow's Open-Ended Method, and yoga for hands); feedback options (including no feedback, directed and minimal feedback, an all-class workshop, a private consult with the instructor or embedded writing center tutor); and revision strategies (including the Montaigne method, "The Corpse Pose," and returning recursively to a prewriting activity) and document their use of these methods in the log entries. Students are encouraged to dwell for as long as possible in nonverbal and prewriting stages and be in no rush to identify a genre or topic; the intent is let students decide the moment that formlessness has sufficiently turned over to form so that they feel the time is right to draft.

The challenge is to assure students that it's perfectly fine to linger in the preverbal, since the openness of this task (in its magnitude of potential distance from polished products) can be unsettling. Student writers often struggle with preconceived standards for what is an acceptable amount of unknowing in a writing process. They associate prewriting and invention with already possessing an idea for a new piece rather than as a time to discover that idea, to occupy a place of formlessness. For other students, the task set-up leads to positive affective responses to writing: "Prewriting again makes me comfortable with where I am at. It brings me to a point where I feel I can write about anything" (Marcus). I point out that this task situation isn't unusual, because any time they are given a new writing assignment a gap occurs between knowing they have to write and knowing what or how they want to write. Prewriting simply gets amped up in this project as students are rewarded for making various metacognitive choices in order to make use of verbal emptiness. About halfway, students submit a "contract" in which they make decisions about content, genre, audience, feedback, and evaluation, with the caveat that they are welcome to revise their contract as they continue to work on the project. One semester, a student's request to change his contract occurred on a near-daily basis in response to changes in his rhetorical situation—audience, topic, and genre were in interplay—resulting in a multi-page prose love poem which he read to the class. Another student adjusted her contract to decrease her proximity to audience: "After doing the contract, I decided that I didn't want anybody in my class to read it. I like the ability to choose my audience" (Claudia). After the entire project's completion, we discuss how it's probably preferable that students did resubmit contracts, since it's indicative of awareness of flux and rhetorical factors.

No-self constitutes an important part of any study of verbal emptiness in the first-year composition classroom. The intent is to help student writers understand how intrapersonal rhetoric installs a misleading writing self. It also teaches how that perception of self often undercuts a richer interconnectivity possible in immersing self in verbal emptiness rather than attempting to make self free-standing and autonomous. No-self teachings walk student writers through the first four of the five *skandhas*, starting with how our intrapersonal rhetoric props up an ego, moving to the evaluative tendencies of the intrapersonal, then to the perception of other discrete entities in the form of readers. In a brief list activity, writers practice dissolving the individual ego by thinking of twenty-five minor and major ways in which they are interconnected to others inside and outside the classroom through their writing—Michel Montaigne for inventing the modern essay, the people who trained their teacher, a single word they've learned this semester and its etymological origins, the classmate who shed new light on paragraph structures, the people who manufactured their computer or notebook, the food and caffeine in their body, or the construction workers and engineers who designed the classroom. Vice versa, students contemplate how they are linked to other students' writing and thinking—an observation made in class or a writing quality of the other person that's left an impression. Through mindful listening while giving and receiving feedback, students divert the normal inclination to sort and react to opinions about their drafts to a new non-evaluative, neutral reception of those opinions. Observing their breathing while listening to or reading suggestions, students evade reactive storylines and instead pay attention to the changing qualities of feedback in real time. Another no-self revision activity involves using another student's suggested wording or idea wholesale and taking note of how this non-self element becomes absorbed into their own draft.

Loving-kindness meditation, an adaptation of Sharon Salzberg's *metta* meditation, is an imaginative practice that shows interconnection and leads to increased compassion for ourselves and other writers and readers, and a non-reactive outlook for writing. What better way to take advantage of the vacancy of the writer's present moment, what better way to make use of our imaginative inclination, than to use those capacities to develop a more reflective, more nuanced relationship to potentially difficult audiences and make connection to more benign audiences? Salzberg suggests that our restlessness, our usual wish for things to be otherwise, is directly connected to our perception that entities are autonomous and therefore either in need or capable of being discretely adjusted. Loving-kindness meditation, on the other hand, lets us "open continuously to the truth of our actual experience, changing our relationship to life" such that people surmount "the illusion of separateness, of not being part of a whole" (21). Students begin by focusing on their inhalation and exhalation. Next, they think of someone who in the past has been supportive of their writing or is currently supportive. On the inhalation, students visualize some event or object which would bring this supportive individual happiness, especially something relating to this individual's writing, reading, or maybe even teaching life. On the exhalation, students visualize this individual receiving this item or experiencing this happy event, like positive publication news or a literary prize. They should continue to contemplate this person in this fashion while watching their breathing for a minute.

In the next step, students contemplate a series of other people: themselves; a "neutral" (a person who is not central to their writing lives but nevertheless plays a role); a tricky audience from their past; any writing group or class with whom the student was involved (such as their composition course or a high-school writing group or a blog); and lastly, all student writers or writers of the genre of their current project. For each type of person, students visualize what would make them happy in terms of their writing. On the inhalation, students visualize this item or event, and again on the exhalation, they visualize the person receiving this item or experiencing this event, and continue this reflection for a minute while watching the breath. From practicing the "Loving-Kindness Meditation for Writers," students often notice a softening of outlook toward difficult readers and a perspective on past readers' responses to their work. Instead of struggling with intrapersonally based audiences, writers can greet that audience with generosity.

4

MIND WAVES, MIND WEEDS, PRECONCEPTIONS

> Just as in a pond of blue or red or white lotuses, some lotuses that are born and grow in the water thrive immersed in the water without rising out of it, and some other lotuses that are born and grow in the water rest on the water's surface, and some other lotuses that are born and grow in the water rise out of the water and stand clear, unwetted by it …
>
> —Ariyapariyesanā Sutra

In Shunryu Suzuki's classic treatise on Zen meditation, *Zen Mind, Beginner's Mind*, two types of mental formations—*mind waves* and *mind weeds*—appear across the surface of consciousness, on the backdrop of emptiness, and throughout open awareness in a metacognition of the present. Mind waves and mind weeds are mental formations that momentarily disturb the calmness of the mind without existing separate from the mind. Resembling a wordless pulse or sensation, a ripple across the surface of emptiness, mind waves usually remain nonverbal because of their brevity—the passing urge to switch the position of one's legs, fleeting irritation at a noise, quick registering of an aftertaste. Mind weeds are more three-dimensional than mind waves because they're discursive, evoking a detail-laden storyline that draws a practitioner away from the moment, and thus they occupy more time in the mind, putting down their roots in the silt of the mind—a daydream that lasts for a minute or a fantasy conversation with a person who is not around. Both mind waves and mind weeds represent opportunities to engage with formlessness, resist binaries, and reach an expansive mind free of divisions of self and other, binaries of positive and negative.

In this chapter, I adapt Suzuki's mind waves and mind weeds to discuss mental formations that contribute to a nondualist understanding of the affective dimension in writing, one that does not rigidly differentiate between logos and pathos, reason and emotion, as part of a Right Effort and Right Attentiveness for composing. The concepts of mind waves and mind weeds provide a spectrum of non-discursive to more discursive affective experiences for composing theory. As constructs, mind waves and mind weeds also supply a present-based, moment-to-moment perspective on affective writing experience that's crucial to maintaining awareness of a present rhetorical moment. I categorize waves and weeds as affect, as forms of underrecognized self-pathos that are both verbal and nonverbal, to highlight their role in intrapersonal rhetoric focused on writing and needing to write. Omitted from conventional theories of writing, these experiences are erroneously deemed irrelevant to the work at hand (producing an essay, presenting research, developing an argument, working in a genre), all the while exerting influence on students' writing. In casual conversations about writing, mind waves might be dismissively labeled "distractions." Of course, denuding student writing of expressed emotion strips away agency, closes access to rhetorical resources, and leads to de-personalized, author-vacated scholarship, as Lisa Langstraat and Jane Tompkins and many others have pointed out, but the emotions generated by and about writing have received less inspection. I focus on emotions generated by the act of writing—affective responses which happen because of engagement with a process or situation—and not emotion expressed as content. Instruction in mindful writing helps student writers recognize the difference between mindful and mindless responses to mind weeds. To increase their control of self-pathos, student writers observe without evaluation mind waves and weeds as they come into view—those red-hot angry lotuses, jealous lotuses, curious lotuses, fearful lotuses, lonely lotuses, self-critical lotuses, proud lotuses, and affectionate lotuses that drift about in all the pulses, sensations, and energies of writing.

The occasion of writing often arouses a host of feelings, some unpleasant (apprehension, doubt, frustration, embarrassment, resentment) and some naturally pleasant (pride, contentment, self-respect, thrill). People are steeped in their own emotions about needing to write. If we teach writing courses, we can probably just look across the classroom to spot negative reactions to the call to write—slumped or protective body language at desks, averted eyes, sighs, and passive aggressive behavior. We may guess from departmental meetings that our colleagues are generally more on the defensive than relaxed about their writing. Even in psychology, a discipline known for the quantity of its faculty productions, the actual median number of faculty who publish is zero (Boice and Jones 567). For most people, hearing about an assignment and its deadline and then needing to start writing, sustain writing, receive and use critique, and finally deal with any consequences of the end results can be stressful even if the individual has a proven track record of success. Intrapersonal rhetoric and embodied responses spike upon learning of a new writing task as pulses increase, stress hormones churn, and preconceptions featuring expected readers flash on the screen of the mind. The occasion of writing is rarely neutral.

Mindful metacognition of the writing moment plays an integral role in the development of a calm mind that remains non-reactive to the parts of our intrapersonal rhetoric that emotionally manipulate our actions and perceptions. The nature of writing emotions necessitates training in metacognition, since as Alice G. Brand says, "the distinctive freedom of affect from attentive control, its speed, and the range and depth of language that results suggest something special about its influence on writing" ("The Why" 441). While we don't select our emotions, their involuntary nature doesn't mitigate our responsibility for them or soften their persuasive effects on us; hence the usefulness of a Buddhist emotional education (Morgan). With the usual fight-or-flight response of future-based thinking, writing frequently becomes an occasion of rebuttal, for being on the defense (against imaginary interlocutors and readers), and a source of stress for some people. Writers often convince themselves of certain limitations of their writing capacities by insentiently manipulating their emotions during self-talk. One intent of mindful composing is to build a long-standing state of satisfaction, calm, a transferrable self-appreciation—a Buddha smile for writing—not bound to a particular assignment (or writing instructor) through metacognitive awareness of these affective arisings. The Buddhist practices of metta or equanimity (calm, non-reactive mindset to internal developments) and maitri (calm, non-reactive mindset to external developments) are forms of clear awareness and ideal ways to encounter weeds and waves—and better manage a rhetorical situation.

Emotional Mindlessness in Composition Theories

For the most part, composition scholars overlook the role of emotions in writing experience, and that's a cause for concern, since it's a contradiction to try to foster metacognition and simultaneously exclude the huge portion of mental activity that emotional formations comprise. In the 1990s, Susan McLeod critiqued cognitivist and social constructivist approaches for their mishandling of affect, maintaining that "[c]ognitivists disallow the writer's subjective experience. So, it seems, do the social constructionists—at least the way it is now envisioned. Emotions seem to be there, but composition theory (and writing courses) apparently shouldn't much bother with it" ("The Affective" 400). Advocating for a "psychology of writing," Alice G. Brand in the 1980s and 1990s claimed that social cognition gave the field "ammunition to avoid studying [students'] emotional experience" ("Social" 396). Of the cognitive model of Linda Flower and John Hayes, Brand points out that their model "provides no language to deal with emotion" and as a result Flower and Hayes' study of cognition falters, lacking as it does the "rich, psychological dynamics of humans in the very act of cognizing" ("The Why" 440). Recently, Amy E. Winans criticized academia for its complicity in "cultivating student illiteracy surrounding emotions" and suggested a move toward what she calls critical emotional literacy (153). The consequences of this denial of writerly emotion show it to be a far cry from Buddhist no-self—it's not a cleaning up of the messes of ego or self-interested expressionism. Rather, such a view elides the interconnectedness of self and other, substituting mindlessness for the interbeing of emptiness, for in trying to downplay student writers' emotions, what paradoxically happens is a diminishment of the social context of those emotions. As Bronwyn T. Williams says, "while our initial response is quick and internal, emotion does not remain limited to our individual, embodied response" but instead is "integrated into, and developed through, our social interactions and contexts" (16). Williams makes a case for the centrality of emotions in mediating writing experiences, pointing out how some students arrive at writing activities with negatively charged memories of writing that impinge upon their agency (16–17). Moreover, these views expect mindless behavior of student writers, since obviating writing emotion means paying less attention to the intrapersonal rhetoric and real-time embodied experiences associated with those emotions. Certainly, the field's disinclination to consider emotions specific to the act of writing seems strange given the unavoidable signs of these emotions in our first-year classrooms—from the body language of students, their underlife grumblings before a class meeting starts, the box of tissues used during office hours, worry about grades and failure of a required composition course, the quiet flush of pride or relief at a task well done and turned in on time.

The second problem with how writing scholarship treats writing emotion is that it frequently perpetuates a dichotomy of reason and emotion, valorizing the former at the expense of the latter. While this is less egregious than a wholesale ignoring of emotion, the field of composition undervalues the rhetorical capacities of emotion by "positioning emotions as outside efforts to reason, communicate and act meaningfully" (Micciche 4). According to Laura Micciche, the discipline's Aristotelian treatment of emotions reduces affect to the merely personal, a move that mirrors the feminization of emotion in popular culture and academia and that runs counter to the integral role held by emotions in everyday life (7). Contemplative pedagogies, however, can assist with the reintegration of emotions. Tobin Hart has said that "[r]ational empiricism trains us to pay attention to some things, and not to others, discounting hunches or feelings, for example, in favor of certain appearances and utility," but that contemplative learning increases understanding by adding "shadowy symbols, feelings, and images ... paradoxes and passions" (37). As I'll shortly explain, while reason and feeling might be locked in a binary in certain factions in composition studies, in Buddhism, mental formations are not categorized as one or the other.

The third problem lies in the lack of particularity in approaches to affect in composition theory, which frequently resorts to atemporal generalizations in which usually negative writing-related emotions are accruals rather than captures of moment-to-moment awareness. Writing affect is often treated as an attitude that is pre-made, brought to the rhetorical scene, rather than manufactured by the student's internal discussion during writing. To borrow Susan McLeod's terminology, these strategies focus more on trait anxiety, or the way in which a student tends to respond to the need to write, rather than state anxiety, or a particular response to a particular writing task ("Working" 97–101). Trait emotions generalize time or habitual responses that "predispose people to particular state emotions but do not have a locus in time" and are atemporal notions which consequentially risk perpetuating mindlessness (Brand qtd. in Musgrove 2).

The typical direction of this line of questioning concerns the degree of interest a student brings to his written work—or attitude. Accordingly, John A. Daly averred, "attitude about writing is just as basic to successful writing as are his or her writing skills," a relevance underscored by Laurence Musgrove's belief that "attention to attitude will offer students the chance to see how what they bring to writing influences what they can ultimately achieve in writing" (Daly 44; Musgrove 2). In his examination of affective attitudinal aspects, Daly narrows the emotional range to writing apprehension or to students' feelings about texts they've already completed (69–70). On the other hand, Daly's differentiation between dispositional attitudes, or habitual feelings about writing that students maintain over time, and situational attitudes, or "transitory feelings closely tied to a particular situation or event," which have received less scholarly scrutiny than dispositional ones, is closer to a perception of the present moment focus of mindful composing. As Daly says, a student's situational attitude occurs as a "feeling that arises in a particular classroom on a certain day while working on a single assignment for one teacher," a time frame that would be further focused on the few seconds of present attention in a mindful approach (66). What we don't cover in composing theories and in first-year classrooms is real-time pathos, or the emotions students feel *as* they write, emotions about needing or wanting to write the project at hand.

Susan McLeod's lexicon of affect, borrowed from cognitive psychology, does much to redress this problem. It considers the duration of these mental formations and differentiates between generalized and particular emotional occasions, thus incorporating the present moment. McLeod divides the broad category of affect into emotion—an embodied response of intensity and brevity, such as grief or ecstasy; moods—responses which are less intense, such as happiness or depression; feelings—not to be confused with emotion, these are embodied sensations that coincide with emotions and moods, such as neck tension or lack of energy; attitude—which is "not a response, but a readiness to respond in certain ways"; anxiety—a condition of disquiet or tension that can be either a trait anxiety (a tendency to respond with anxiety) or state anxiety (specific to a particular event or task); beliefs and belief systems—inferences that are connected with expected outcomes; and motivation, both intrinsic and extrinsic, physiological or psychological, or the affective condition that directs behavior ("Working" 97–101).

An example of affect research which partially adopts a moment-specific approach is John Daly's work on writing apprehension. Daly contrasts the moment-specific category of situational attitudes, which he defines as "transitory feelings closely tied to a particular situation or event," and dispositional attitudes, or generalized and persistent feelings (65–66). The actions behind situational attitudes suggest meta-awareness, since they entail "perceptions of writing contexts, responses to the writing situation, and perceived outcomes" (71). He identifies variables contributing to that perception of context as an emotional, critically aware activity as the "degree of evaluation perceived present in the setting," "the amount of task ambiguity," "the degree of conspicuousness felt," "the perceived level of task difficulty," "the amount of prior experience writers feel they have for a task," "the personal salience" of the task, "the degree of novelty attached to both setting and task," and the students' anticipation of their readers' responses (72). However, Daly's attentiveness to the fluctuations of context is subsequently undercut by his lengthy discussion of the existence of dispositional attitudes about writing. According to Daly, writers experience "relatively enduring tendencies to like or dislike, approach or avoid, enjoy or fear writing" and are "assumed to behave in a more or less consistent manner when it comes to writing" (44). In categorizing people as more or less writing apprehensive, he omits their present rhetorical context and assumes a static, fixed writing self.

Finally, the range of possible writing-related emotions is pruned to anxiety, apprehension, and sometimes boredom in the scholarship on writerly affect. So many nuanced writing emotions are omitted, not to mention their moment-to-moment gradations. It's reminiscent of how a limited emotional range was evident in Aristotle's coverage of pathos in *Rhetoric*—mainly anger (broken down into different types of slights) and its supposed opposite, calm. What is needed from a study of self-pathos is a more microscopic, fine-tuned perception of the full range of emotional responses. As odd as it is to have to say this, it's important that the range include first-year composition students' positive feelings about writing, since there's near-complete radio silence on this area. In one case, the satisfaction a student feels after completing a draft is registered as a problematic resistance to revision, and a response to be tamped down by teacherly intervention (McLeod "Some"). Nathan Crick avows: "Success in writing will then be judged not by whether [students] accurately represent their thoughts or resist dominant discourses but whether the words they create inspire themselves and those around them to experience the joy of Becoming in the midst of their own writing" (273). As will be discussed, writing emotions should be approached with Buddhist non-discrimination: all emotions are welcome, none discarded. It's the more momentary and more subtle emotions of mind weeds and waves that put a student writer and the teacher's pedagogy at the crossroads. Depending on how weeds and waves are handled, suffering or release from suffering ensues, clutching at pleasure or a non-attached experience of pleasure ensues, and for students, the quality of their writing experiences is determined by how they handle emotional mental formations by observing rather than suppressing unwelcome constraints.

The Buddhist Mindfulness Approach to Emotions

Buddhist practices offer instruction in the perception, regulation, and acceptance of the emotions, which is applicable to affective experiences in first-year writing pedagogy. In the Mahāyāna school of Buddhism, the educational model for the emotions is both normative and pedagogical—normative in that it installs certain emotional behaviors as meritorious (calm, compassion) while others are deemed destructive (anger, envy, lust) and pedagogical because it's teachable and not based on innate character qualities (Morgan 291–292). Before we enter into the details of meritorious emotions, it's important to point out the complex balancing act Buddhist mindfulness performs between designating certain emotions as positive and continuing to practice detachment through non-evaluation. To cherry pick between affective experiences could easily seem hypocritical: wouldn't that indicate that an individual is clinging to one sort of phenomenon and thereby resisting change and the interconnection of that phenomenon with others? After all, the whole enterprise of Buddhism is to avoid a single mega-emotion through radical non-avoidance—suffering! Actually, the mindfulness approach to emotions means perceiving arising emotions (of any ilk) but not slavishly following them around by letting our mind slip into reactive storylines. Thich Nhat Hanh explains that "[m]indfulness does not fight anger or despair. Mindfulness is there in order to recognize ... This is not an act of suppression or fighting" (*Anger* 164). In fact, an added challenge is our tendency to avoid unpleasant emotions (and seek out pleasant ones). As Pema Chödrön says, "[g]enerally speaking, we regard discomfort in any form as bad news" but unpleasant emotions such as disappointment or fear "teach us to perk up and lean in when we feel we'd rather collapse and back away" (*When* 12). Along similar lines, the goal in vipassana meditation is to perceive such mental formations as they arise with detachment, to "[l]ook on all of it as equal, and [to] make yourself comfortable with whatever happens" (Gunaratana 40). Receiving a rejection notification, a poor grade, or agitating criticism are all opportunities to abide in the moment and observe emotional fluctuations without feeling compelled to reject or become mesmerized by them.

With a mindful composing pedagogy, we seek to reduce the writing suffering of our students, but at times this means perceiving but not rejecting suffering, from minor discomfort to more serious stress, on a backdrop of metacognitive calm. The paradox is that suffering is categorized as a negative phenomenon, yet part of the method for finding relief is not to overlook difficult emotions. As Paul Ekman and his co-authors explain the two levels of Buddhist practice with emotions, the first is to "identify how [destructive mental states] arise, how they are experienced, and how they influence oneself and other over the long run" and at the next level, a practitioner "learns to transform and finally free oneself from all afflictive states" (60). Writers should sit with difficult writing-related emotions, watching the emotions almost like a fire in a fireplace for their fluctuations in energy and imagery, without succumbing to the temptation of running off into the emotion's storylines. Essentially, this practice entails noticing an emotion without judgment while simultaneously remaining aware of one's present circumstances. This discipline also means teaching writers not to cling to even positive feelings about writing, because peak emotional experiences such as pride or excitement warrant mindful monitoring to curtail preconception and storyline.

Mental formations are called *kleśa* or afflictions, alternatively "taints," "hindrances," and more vividly poisons, and the point of observing *kleśa* is to mitigate mindless reactions which perpetuate karma, harming self and other. Donald Lopez explains that these are "states of mind that motivate the performance of the nonvirtuous deeds that in turn produce suffering" and that Buddhist treatises go into great depth in connecting specific afflictions to specific nonvirtuous deeds (46). The education in the emotions begins with the Four Noble Truths, in which the "first three Truths articulate a norm for emotional life, including an ideal (*nirvana*) along with a diagnosis of how we fall short of that ideal (*dukkha*) and an account of the cause of falling short of the ideal (craving)" (Morgan 296). In addition, Right Effort features four practices pertaining to arising mental formations: to avoid them (through a control of the senses, the meditator steers clear of attachments which might arise); to overcome them (devotion of mental energy to dealing with attachments which have already arisen); to develop them (through a control of the senses, the meditator steers toward positive attachments which could arise); and to maintain them (devotion of mental energy to grow positive formations that have already arisen) (Goddard 45–47). With Right Effort, the sorts of negative formations a practitioner avoids are "evil and demeritorious things" such as lust, greed, anger, and ill-will, while positive mental formations to promote are ones that lead to enlightenment, such as feelings concerning solitude, detachment, attentiveness, tranquility, and equanimity (Goddard 46).

Although reason and feeling might be locked in a binary in certain factions in composition studies, in Buddhism, mental formations are not categorized as one or the other. For instance, in the several ancient languages of Buddhist texts, there is no single word for general emotion but, instead, words for specific emotional states (De Silva 109; Ekman et al. 59). Furthermore, in the Sammādiṭṭhi Sutra or "Right View," six classes of feeling are enumerated in a blend of what we would call thinking, physical sensation, and emotion: "feeling born of eye-contact, feeling born of ear-contact, feeling born of nose-contact, feeling born of tongue-contact, feeling born of body-contact, feeling born of mind-contact" (Ñāṇamolí and Bodhi 139). Neuroscience maintains this nondualism, since neurological systems which support emotions and thought are interwoven—an anatomical manifestation of nondualism (Ekman et al. 59). Likewise, an approach in composition theory that adopts mind waves and weeds might be equated with the "incredible, dynamic use of energy that we rather nonchalantly encapsulate in the word *thinking*" (Pullen 24). In discussing mind waves and mind weeds, Suzuki uses the catch-all phrase "thought," despite an apparent distance of these impulsive, fleeting mental formations from what might be considered logos. Furthermore, the main distinction Suzuki draws between waves and weeds is essentially the duration of their occurrence, the former briefer and less elaborate than the latter, despite how mind waves, as near-wordless to fully wordless formations, may seem very far from logos. I suspect that references to thought in discussions of writing are code for whatever discursive material seems audience appropriate, geared toward eventual public display. Attention to mind waves can especially help student writers gain distance from excessive logocentricism, since waves are less discursive than weeds (and less resemble any eventual textual production).

In "Renegade Emotion: Buddhist Precedents for Returning Rationality to the Heart," Peter D. Hershock contends that Buddhism doesn't distinguish between reason and emotions; to do so would be a refutation of emptiness and interdependence. Emotions are aligned with the interconnectivity of emptiness whereby emotions become "relational transformations through which the direction and qualitative intensities of our interdependence are situationally negotiated, enhanced, and revised" (252). Hershock talks of a "dramatic partnership" between all entities, similar to Thich Nhat Hanh's interbeing, which takes place in the middle ground—or emptiness—between those entities. It becomes an improvisational act, this development of moment-by-moment connection, of "continuously negotiated shifts in the meaning of our situation away from *samsāra* and toward *nirvāna*" (254). Adopting a view toward no-self, Hershock says emotions are not experienced by an abiding self but, rather, occur on that middle ground and are the results of "situational negotiations of the intensity and direction of our dramatic interdependence" (255). This kairotic negotiation speaks to Laura Micciche's notion of "emotion as emerging relationally, in encounters between people, so that emotion takes form *between* bodies rather than residing *in* them" (13); so the rush of blood to the face, elevated pulse, and flurry of anxious imagery of embarrassment are not fenced inside the single chagrined person. Hershock particularly takes a critical look at the brand of reason which denies the middle ground between existence and non-existence by insisting on firm categories and identities, by implication including the binary of emotion-reason (252–253). Static entities are replaced by ones defined by their interrelatedness to other entities, defined by their connections to a larger context (253). Karma is in the house whenever a person ignores this interconnectivity and binds other entities and phenomena to his or her own "horizons for relevance," a cognitive place in which the drama of liking and disliking plays out, causing the cycle of suffering (253). Reason becomes reliant on emotion in this account, because reason can only take us so far in a mindfulness perspective: reason will need to rely on the relational dynamic of emptiness, that connectivity of emotion (257). Ultimately, his project is to alter the reason/emotion binary (in which the latter often receives less import) so that reason constitutes a type of emotion.

Mind Waves and Mind Weeds in Students' Experiences of Writing

Suzuki's constructs of mind waves and mind weeds contribute to composing theory a nondualist and present-based understanding of the affective dimension in composing, one that avoids differentiation between reason and emotion. As ways of perceiving the emotions of writing, mind waves and weeds subtly put into interplay mental activities less possible with larger labels like "anxiety" or "apprehension," making possible a pedagogy that "recognizes the unquantifiable aspects of meaning-making: intuition, values, felt sense, moods, emotions" and provides "access to these slippery areas" (Fleckenstein *Embodied* 26; 28). In Suzuki's explanation of these mental formations, there isn't a parsing of formations that arise from thought and others from feeling. Instead, it's all an "expression of big mind," a consciousness that does not distinguish between self and other (*Zen* 35). By including non-discursive and often fleeting affect, mind waves and mind weeds offer a nuanced view into intrapersonal dynamic—a more fine-tuned notion of affect than categories such as emotion, feeling, or even intuition. Mind waves and weeds, especially waves, resemble the "subtle energies," the tracking of which Terri G. Pullen says leads to focus, an "active receptivity," and an all-in-all "positive, more energy-efficient mind set" that reduces trepidation in composition classrooms (24–25; 29). So mind waves that pertain to writing might include irritation, envy, doubt, a pulse emitted by the connotation of a word or memory, the adrenalin flash of realizing that one is running out of time to finish, exhaustion, swoosh of pride, a sudden pain in one's wrist, bolt of fear, strike of hunger or fog of low blood sugar, tug of guilt, a passing question about worth or purpose, any accumulated emotional build-up from previous writing experiences, the prickle of self-criticism, the thrill of exceeding a word count, the satisfaction of hearing one's rapid typing.

More three-dimensional and discursive, mind weeds pertaining to writing might mean imagining a roommate reading a current draft; wondering if so much time writing is ruining or benefiting one's social well-being/physical health/mental health; fantasizing about how one will reward oneself when the writing project is successfully completed; comparing one's current writing abilities with one's previous writing abilities or those of another person; dreading reaching the next phase of the project; visualizing oneself published or getting an A (or F); thinking one should be doing some task other than writing; imagining other available research in a database; seeing oneself repeatedly interrupted; thinking that a deadline is impossible; expecting to use a past strategy; assuming that one will remain within a given genre during a writing session; setting up a stylistic rule, such as that one must use short sentences. Frequently non-descript, elliptical, and undramatic, waves and weeds—but especially waves—are the contents of a passing moment, their transient quality precluding them from becoming writing blocks or other composing problems.

Students should not be bothered by their mind waves or weeds, not judge the emotions discovered in their writing minds, since mind waves and mind weeds will dissipate and other formations will emerge. These formations are natural, inevitable, and not worthy of evaluation, just as we wouldn't try to isolate a passing wave from its pond or feel critical of the pond for including waves. Mindful meta-cognition circumvents our tendency to evaluate emotional experiences as either pleasant, neutral, or unpleasant, as the Buddha explained for Right Attentiveness:

> In experiencing feelings, the disciple knows: I have an agreeable feeling, or: I have a disagreeable feeling, or: I have an indifferent feeling; or: I have a worldly agreeable feeling, or: I have an unworldly agreeable feeling; or: I have a worldly disagreeable feeling, or: I have an unworldly disagreeable feeling; or: I have a worldly indifferent feeling, or: I have an unworldly indifferent feeling.
>
> *(Goddard 50)*

Mind waves are not separate from the mind: to conceive of writing emotions as such would be to perpetuate a false dualism. As Suzuki says, "A mind with waves in it is not a disturbed mind, but actually an amplified one" and "[a]ctually water always has waves. Waves the practice of the water" (35). Similarly, mind weeds are like cognitive dandelions: they invariably happen, but the important point is that weeds can be pulled out, examined, and used to fertilize mindfulness: "[w]e pull the weeds and bury them near the plant to give it nourishment ... So you should not be bothered by your mind. You should rather be grateful for the weeds, because eventually they will enrich your practice" (36). It's not self-pathos per se that leads to composing problems but, rather, our responses to those emotions, and specifically how we categorize or allow them to swell into storyline preconceptions in what Terri Pullen describes as "the over-extension of these categories" (25).

Instead of resisting affective formations, to experience calm mind, a practitioner is advised to watch his or her breathing and develop an appreciation of these interruptions in a practice of "no effort" in which struggle and evaluation are not added onto the moment. Suzuki's take on effort is no effort, or a nonconventional type of trying: we're not to try to stop our mind, just observe its transience; as he says, "do not try to stop your thinking. Let it stop by itself. If something comes into your mind, let it come in, and let it go out. It will not stay long" (34). From a mindfulness perspective, it is important to observe rather than block out even extremely difficult emotions, registering when they arise. Along these lines, in contemporary Buddhism, fear and anger receive substantial coverage, with practices entailing a clear-seeing exploration of the roots of such feelings in order to return to open awareness (Chödrön *When* 1–5; Hanh *Anger* 27–28; Trungpa *Meditation* 48). In the first-year writing classroom, difficult emotions should be seen as part of the writing occasion, rather than denied or resisted, and "we can tell students that all writers are agitated as they compose, and that they can learn to find that agitation enabling rather than debilitating" (McLeod "Some" 433). The need to abide by writing discomfort is crucial, because as Roger Bruning and Christy Horn have suggested, "[a]s affect turns negative, the natural consequence is that students will begin to avoid writing wherever possible," a cycle which can impact educational major and career choice (33; Daly and Shamo). Consequently, any writing emotion is acceptable—no matter how seemingly shameful, childish, embarrassing, malicious, selfish, egotistical, or disloyal. From a mindful composing practice, the whole array of human emotions can pass through a mind without suggesting anything fixed about the individual's personality, nature, or prospects. A moment of defeat or laziness can occur a few seconds before another affective pulse of generosity or confidence, and vice versa, the seemingly positive will be followed by the seemingly negative. The emotional constraints of composing are always in flux, such that we should consider "the affective domain not as products or outcomes ... but as dynamic states of being" (McLeod "The Affective" 102). Writing emotions are not static constants that can be separated from the context of the moment but are dependently arising formations interwoven with emptiness: as a result, weeds and waves are not indicators of a student's fixed ability.

When students mindfully write, they strive not to add anything extra to the experience, including extra thoughts of value or evaluation. Instead, students learn to observe with a lightly detached mindset the contents of each moment, both for possible material for drafts and for metacognitive information about factors in the rhetorical situation. Writing-related emotions should not be evaluated as they manifest: the student writer should be steered from thoughts such as *I shouldn't feel irritated that I have this assignment* or *it's so stupid that I have all these doubts still about my writing.* Suzuki's advice about waves is that we don't become concerned about their arrival (concern only adds more waves) and instead just observe their arrival and departure: "if you are not bothered by the waves, gradually they will become calmer and calmer. In five or at most ten minutes, your mind will be completely serene and calm" (*Zen* 34). For writing students, adding something extra might mean a thought about one's virtuousness in starting a project early or doubt about completing by deadline. Adding something extra involves real-time judgments, those internal critiques of emerging text which often cause writing students to jumble composing with editing, creating with correcting. This addition might entail thoughts of comparison—comparing a current writing experience with a prior or comparing one's writing ability with that of another student, friend, author, or relative. As Suzuki says, "When you do something, you should burn yourself completely, like a good bonfire, leaving no trace of yourself" (62). This strategy is similar to the "not trying" described by Rachel Forrester, who finds value in a "seeming absence of human effort" in rhetorical silence and in desisting from analysis and reason while writing (46–47). It is too tempting to construct elaborate storylines around these extra tidbits: "If you leave a trace of your thinking on your activity, you will be attached to the trace" (Suzuki 62). Mental formations of ambition, excitement, failure, or disappointment are extraneous material which lessens a conscious experience of writing. Even the thought that to write is a positive phenomenon and to not write is a negative phenomenon need to be dropped and replaced with a radical ambivalence toward the act of writing. It is neither commendable nor encouraging that writers write; likewise, it is neither blameworthy nor discouraging that writers are not writing.

Establishing a Calm Outlook for Writing

In mindful first-year composition, students actively work to change their affective responses to writing occasions, learning not just to observe those affective responses but also to intentionally alter their self-rhetoric. Certainly, teachers should strive to add positive associations of enjoyment and fun to students' repertoire of writing memories, but students should also take matters in their own hands and endeavor to alter their self-pathos such that the dominant experience is of calm. This mindful rhetorical awareness calls for proactivity from student writers, since, as Robert Boice put it, "fluent writers make their own moods rather than wait for them" ("Writerly" 34). Mindful metacognition is useful for a range of observations—noticing intrapersonal rhetoric for possible content, noticing the physical absence of audience, and so forth. This metacognition or bare awareness pertains directly to emotional experiences and at the same time carries its own affective attribute of tranquility. A sensation of attentive, open-hearted acceptance becomes the replacement emotion for the general apprehension many students experience with writing. In other words, calm is what results after observation of and detachment from writing-related, ever-arising emotions. It should be repeated that the pursuit of this state of calm is not intended to obliterate other emotions that arise: rather, calm and balance serve as the overriding or background emotion of the student, a template upon which the other emotions rise and fall. In essence, writing calm is the self-pathos of a metacognitive mind. Aristotle situated calm as a form of pathos in *Rhetoric* by casting it in a binary relation to anger, saying that "[G]rowing calm may be defined as a settling down or quieting of anger" (216). Calm, however, from a mindfulness perspective is not the opposite of troubling emotion but, rather, the psycho-physical response to internal rhetoric that is able to treat arising emotions with a combination of awareness and detachment, part of the "relaxed, self-reflective pedagogy" that Geraldine DeLuca describes (31). Although the occasion of writing is rarely neutral in practices outside of mindfulness, through mindfulness, I want to make writing a neutral experience: a gray rock of ambivalence balanced in a kairn.

With mindful calm, pronounced dislike or writing is not warranted, and neither is pronounced enthusiasm for writing, since ultimately a mindful writer seeks ambivalence about the ability to write and writing successes. In *How Writers Journey to Comfort and Fluency*, Robert Boice advised faculty writers to seek a "mild happiness" or a "wise passiveness" in writing and avoid binging or forcing, underscoring the "value of serenity and planfulness in writing"—advice useful to first-year students (4;30). In addition to establishing an all-around pleasant learning environment, mild happiness carries rhetorical benefits because it "broadens the range and flexibility of our associations in thinking," akin to Ellen Langer's idea that a mindful outlook increases the number of perceived options and perspectives (Boice *How* 157). To prevent hypomania and burn-out, Boice recommended frequent stops during writing activity to prevent an overvaluation of writing at the expense of other life activities (*How* 33). To overly appraise a writing experience or outcome is to add something extra to consciousness—leading to a ripple of mind waves or the growth of a mind weed which will then need mindful attention. It can be hard to temper one's reaction to a change in writing ability from struggle to flow, but the ability to remain mindfully calm requires that writers not overly value writing and ultimately accept a less productive writing session equally with a productive one, bowing to difficulty with the same gratitude as one might bow to success. For example, Reed Larson mentions a student who is able to note the vicissitudes in his writing experience with remarkable equanimity and self-care, modulating his writing process. This student "put in a lot of effort and closely monitored his energy level" and "seems to have been deliberately adjusting the challenges to his abilities" (35). As a result, "By moving cautiously through hard parts, by stopping himself when overexcited, and by monitoring his energy, he regulated the balance of challenges and skills, creating conditions for enjoyable involvement" (35). The student's ability to monitor his "internal states protected him from the kind of overexcitement and anxiety that paralyzed" classmates (35). What's notable is how the student, by observing rather than overlooking fluctuations in his affective responses to writing, essentially reaches a condition of balance, characteristic of mindfulness. Observing emotions in this way subjects them to flux, and they become more detailed (constant change necessarily multiplies details and kicks up more information) and ego is less rigid.

The legendary taciturnity of the Buddha symbolizes an outlook of equanimity that is possible for writing through mindful metacognition. We won't find sculptures or paintings of the Buddha waving a sword in battle, eating a meal, or even teaching in a conventional lecture pose (the Buddha was said to occasionally meditate straight through his instructional sessions). To illustrate calm versus troubled affect in writing, I show my first-year composition students three iconoclastic images of men in the midst of silent contemplation, two of whom are clearly suffering because of their relation to their thoughts. In all its twenty-eight versions, Rodin's "The Thinker" looks stark and uncomfortable (the figure is naked, leaning forward a bit precariously on a stump-like outcrop, face resting on a hand which is resting on one leg, frown line in forehead) and appears as though he is tracking a difficult line of thought. Whatever he's thinking about, the activity isn't a relaxing one for Rodin's thinker—not coincidentally, the same figure in miniature appears in Rodin's "The Gates of Hell." Thomas Eakins' painting "The Thinker: Portrait of Louis N. Kenton," his brother-in-law, in contrast, is of a brooding melancholic, a full-length portrait of a man dressed all in black, his hands in his pockets, his shoulders hunched, legs slightly akimbo, his gaze directed at the floor, his mouth slightly downturned. Kenton stands alone on the backdrop of the canvas, a separateness emphasized by the way the majority of his body is in contrastive black. The image of a seated meditating Buddha is as iconic of Buddhism as the crucifixion is for Christ; the selection of representative images for these religions is telling, one an image of how to cogitate and the other a horrific image of self-sacrifice in a sadistic execution practice routinely used by the ancient Romans. With the large outdoor sculpture of the Buddha in Kamakura, Japan, the figure's face is relaxed, shoulders softly lowered, eyes heavy lidded, and most strikingly, a continuous flow of movement is suggested in this inert figure from the loop of the Buddha's robes across his chest. His hands are positioned in a throat chakra or mudra dhyana, the thumbs and fingers touching in a symbol of balance, a bodily representation that mirrors a mind able to observe continuously passing thought with detachment that prevents vulnerability to bondage to a storyline or vulnerability to fumes of mood.

Writers can analyze their own intrapersonal rhetoric to look for factors including self-pathos or how they convince themselves through emotions. Students' self-talk during writing exerts a powerful self-pathos throughout an ongoing inner discussion about writing process, aspects of the rhetorical situation, and emerging content, and that self-pathos in turn affects those three matters. Self-talk during composing transmits puissant messages through what Robert Boice has identified as the seven types of "pathological thinking that accompany writing blocks"; top of the list is writing apprehension, followed by impatience and anxiety over the prospect of rejection ("Cognitive Components" 91; 100). It bears repeating that the material of the intrapersonal is never an identical copy of the piece that is interpersonally shared with a reader. Instead, the intrapersonal is about the type of experience a student privately gives herself, how she talks about her ongoing and prior writing experiences, immediate and previous writing performances—what sort of narrator or guide she is to her own writing process. As Laurence E. Musgrove points out, "What is within writers is quite different from what they've produced, or even the knowledge and skills they've developed along the way" (1). Emotions displayed to the self are rhetorical and play a role in "framing the context of decision making," and those displayed to the writing self are no exception (Feleppa 263). Calm can be induced psychosomatically through breath work (or even through as basic a gesture as lowering the tongue to the floor of the mouth, as Boice describes), but for a deeper analytic awareness of internal rhetoric, formal analysis assignments are particularly effective (*How* 29–30). In the previous chapter, we discussed self-rhetorical analysis assignments as a way to better perceive and manage intrapersonal interactions; this assignment can be adjusted to focus exclusively on self-pathos (and preconceptions) or include it as a factor for analysis. Just as writers sway readers' emotions through connotation, punctuation, rhythm of sentences, imagery, and other devices, students sway their own views and decisions about their writing through diction, repetition, imagery, and rhetorical questions, to name a few devices, all more or less shaped, more or less verbalized, more or less voluntary or involuntary. Likewise, much as first-year composition students can increase their rhetorical awareness through a rhetorical analysis of a published speech or essay, they can perform internal rhetorical analysis to build metacognitive skills important to navigating the writing process and managing a rhetorical situation.

Two Buddhist calming practices which can be adopted as introductory rhetorical strategies for dealing with affect are *maitri* and *metta*. Maitri refers to a stance of radical, gentle acceptance of any wave or weed arising in one's consciousness, a "befriending who we are already," involving self-kindness, accurate perception, and detachment (Chödrön, *The Wisdom* 4–5). Maitri counters our default response of self-criticism after noticing a mind wave or weed: it addresses a writer's stance on internally arising events. Metta or equanimity practice is a strategy for external events and for installing a large space around observations to avoid becoming reactive and running off into storyline (Sharon Salzberg). As Pema Chödrön notes, individuals vary as to the seriousness of the event that sets off their affective responses: for some, it will take a tragedy, while for others, a house plant moved to a different corner of the house may trigger mindlessness and necessitate these affective practices. So for one student, a failing grade would be a serious event, while for another, a suggestion on a word choice could cause self-talk leading to a consequential affective response. Loving-kindness meditation, as discussed in the previous chapter, can develop appreciation of the interconnection of emptiness, but it also assists with writing affect. This sort of imaginative, empathetic consideration of self and others is evident in early Buddhist writings in which a practitioner "dwells in contemplation of the feelings, either with regard to his own person, or to other persons, or to both" (Goddard 50). Another result of loving-kindness meditation work is that it develops emotional flexibility and resilience, because a "loving mind can observe joy and peace in one moment, and then grief in the next moment, and it will not be shattered by the change" (Salzberg 23). For composition students, loving-kindness meditation potentially impacts students' intrapersonal rhetoric about writing, stocking that inner conversation with more positive, accepting, and generous self-pathos and discourse that changes the direction of their writing.

Preconceptions about Writing Task and Ability

The most pernicious type of mind weed for writing is that of the preconception, an elaborate storyline that makes predictions about the writing task and overlooks the actual content of the present rhetorical situation. Preconceptions are false attempts to control impermanence—a type of grasping—and can take many forms, including rigid rules for composing, fixed ideas about revision, and assumptions about audience and self-efficacy. Preconceptions correspond to what Ellen Langer in her research on mindfulness and learning calls "premature cognitive commitments" and "entrapment by category" in which individuals are "committed to one predetermined use of information, and other possible uses or applications are not explored" (*Mindfulness* 22). By sheltering preconceived ideas, a writer substitutes one form of unknowing, that of groundlessness and verbal emptiness, with another form of unknowing, that of preconception, which is of a far lower grade. So there are preconceptions of the word count that would make a writing session productive, preconceptions about desirable content and genre, preconceptions about how long a writing process will take, preconceptions about audience, preconceptions about organization, preconceptions about knowledge and ability, and even preconceptions about preconceptions. In a nutshell, preconceptions represent a rigid, rules-bound, categorical, acontextual approach to experience that blinds a person to what Langer and Alison I. Piper call the "novelty" of an occasion.

As an assumption conveyed through intrapersonal rhetoric, preconceptions result from a wrong-headed replacement of verbal emptiness, that backdrop of interconnection and inventive possibility, with best guesses that have consequence. Overlooked writing emotion can easily expand into a full-blown preconception about writing task or general writing ability. A seemingly negative emotion arising in one moment not only affects the immediate composing choices of the oblivious student but can also magnify into substantial assumptions about future writing endeavors. In Suzuki's account of preconceptions, "[B]efore we act we think, and this thinking leaves some trace. Our activity is shadowed by some preconceived idea. The thinking not only leaves some trace or shadow, but also gives us many other notions about other activities and things" (62). Premature cognitive commitments accrue over time and instantaneously occur, such that even a temporary decision to decline to critically examine a present moment can send a person on a trajectory of mindlessness as much as a long-term habit (Chanowitz and Langer 1052). In Langer's view, preconceived thinking harms students' self-efficacy, engagement, and ability to identify options and resources (*Mindfulness* 129). Overlearning can even lead to self-doubt, says Langer, since competence that's built upon rote learning may too closely resemble incompetence: mindlessness can make us lose track of our knowledge (*Mindfulness* 20). We lose conscious control of a story we're telling ourselves.

A preconception forms when a writer latches onto a single mind weed concerning a writing task, allowing it to lengthen into a full-blown storyline through details and plot lines, separating the weed from its context of constant change. The mental formation becomes a discrete, autonomous phenomenon—removed from its backdrop of interconnection and emptiness—and a potential source for writing suffering. Robert Boice's opinion was that "[t]he worse the writer, the greater the inattention to emotions while writing" due to the role awareness of writing emotion plays in solving creative-rhetorical problems, reviewing one's work, managing audience expectations, and finding motivation (*How* 159). Similar to a rigid composing rule, except more overtly self-evaluative and built from self-pathos, an emotionally charged thought about how to proceed with a structural component or about the student's supposed ability (probably a combination of the two) is carried into a subsequent moment, impacting the student's decisions: *I excel at revision, I'm terrible at conclusions or transitions*, and so on. It's important to note that a preconception might not necessarily be disparaging, since a writer can maintain positive assumptions about an upcoming writing moment, which may nevertheless diminish the writing act because of mindlessness. Occasionally, a mind wave gathers enough momentum in an individual's discursive thinking to transform into a more three-dimensional and imagistic mind weed, which evolves into a preconception—so a brief sensation triggers a more nuanced image, which lengthens into a full assumption.

It seems, too, that the majority of student writers tend to be attached to negative or self-critical impulses about their writing rather than positive ones. Part of the problem lies in the way a writer's focus is subsequently removed from the present rhetorical moment—that zone of possibility—for the poorer option of a moment stripped of context and saturated in defensiveness. Certain of these mind weeds turned into preconceptions are reoccurring because of their persuasive power over the individual writer: typically ones based around past performance and evaluation. These preconceptions impact a student's self-ethos as a writer or the way in which he positions his credibility, reliability, and effectiveness to himself in ongoing intrapersonal rhetoric, in a line of thinking that usually goes something like *I'm this kind of writer, I'm really bad at this genre*, or *I always impress teachers with my conclusions.* A perspective on Buddhist emptiness can offer students relief from the tyranny of seemingly positive or negative emotions by underscoring how emotions "like all other events, are caused by conditions and will therefore pass in time" and that emotions can be experienced "without exaggerating their importance" (Morgan 298; 302). Each summer I witness first-hand how students make large educational decisions based on their stories of ability during the online writing self-placement program at my university in which students make a case for their first-year composition selection by discussing their previous writing experiences in a writing sample. Frequently, a single past incident with an assignment or teacher shapes the student's self-efficacy: it is superimposed years later onto a different moment and continues to persuade the writer with its pathos.

Mindfully observed as impermanent manifestations, mind waves and weeds are less likely to warp into preconceptions. Neither waves nor weeds can be anticipated in advance, since they're not stable traits a student brings to the present rhetorical situation. Mind waves and weeds occupy unique and ever-changing moments in time. A splash of dread, pleasure at a good-fitting word, or a few seconds of a daydream pertaining to verbal achievement are not affective responses a student can premeditate or import into the writing occasion. As the third component of a Buddhist education in emotions, Right Mindfulness, the second step on the Eightfold Noble Path, helps people recognize emotions as just one more ever-changing event to be observed without running off into their storyline and fantasy (Morgan 300). Watch any writing emotion long enough and its ongoing changes will be evident, just as a fire in a fireplace never stays the same. Much is at stake in checking emotions about writing before they become longer-lasting preconceptions. Debilitating writing blocks happen when a student repeats a preconception as part of her self-rhetoric about writing, dwelling on a stale storyline and failing to notice the offerings of the rhetorical moment. Upon mindful inspection, a writing block reveals a host of sensations, some of them surprisingly lighter, more positive, suggesting a break in the problem. A student may discover a subtle mix of curiosity and trepidation in her affective response to a revision task, and a seasoned writer might note how every third thought about a section in his manuscript presupposes an outcome. It is imperative that first-year students look for the gradations in affect in the moment so as not to cast writing emotions as monolithic and abiding.

Amid red-hot angry lotuses, jealous lotuses, curious lotuses, fearful lotuses, lonely lotuses, self-critical lotuses, and proud lotuses, amid the pulses, sensations, and energies of writing, the mindful writer sits. In a new kind of writerly presence, the mindful student reposes at the center of her activity in a watchful calm that is interrupted by peaks and spikes, by splashes and ripples across the surface of consciousness, but that calmly subsides. More enactive than happiness, anger, or sadness, which are like lightbulbs merely emitting watts of feeling, calm is the by-product as well as the producer of metacognition—it's *calming* as well as *calm*. Amid this placidity, the student has far less reason to stop writing, because the sensation of calm is an intrinsic reward, and for mindful writers, existing in this affective condition replaces word count, grades, and other external markers of success. The mindful writer is a different person from the harried, disinterested, dissatisfied person who was tempted to check text messages or see what others are up to in the dorm. This writer is certainly a different student from the one who was inclined to quit trying, who was drowning in inadequacy and worry. Calm is the emotion of metacognition—the pleasure of metacognition.

Interchapter 4

Detachment, Neutrality, Ambivalence

Detachment, neutrality, ambivalence: none of the goals I have in teaching of mindful affect are goals typically desired by educators. Ostensibly, these are qualities the majority of teachers work hard to fend off as forms of disengagement and boredom. As I hope has become clear from the previous chapter, the ambivalence I seek is different from the usual conception. Far from boredom or apathy, this neutrality is a mental balance that comes with conscious work and critical perspective, arising from recognizing rather than denying the range of emotions students regularly experience because of writing, and then returning to bare awareness. It's a metacognitive calm that results from a clearer understanding of the role of writing-related emotions in shaping composing experiences, outcomes, and self-efficacy. Student writers are awash with often potent emotions about the need to write; when this aspect of their rhetorical situation is repressed, numerous occasions of self-pathos will continue to influence their writing behind the scenes, outside any monitoring.

Course assignments and activities provide practice in bare awareness in order to perceive the fleeting and pulse-like occasions of mind waves along with more three-dimensional mind weeds and accompanying storylines. To that end, some exercises are designed to heighten attention to these emotions, while others are geared toward gaining critical distance through bare awareness, which in turn leads to another affective response—an overriding calm. Together, they comprise a Right Effort of writing in the handling of arising emotional formations: steering clear of troubling emotions through a control of the senses; diminishing the amount of mental energy expended on writing emotions; developing positive writing emotions; and sustaining those positive writing emotions.

It's not as though the mindful writing classroom environment is devoid of feeling. I hope that my students enjoy writing more, deriving professional and personal pleasure from it, and that the build-up of negative assumptions from past moments is scrubbed away by positive ones. In this sense, I am advocating a sorting of emotions, replacing negative memories with more pleasant ones—even crescendos of excitement and thrill like news of a publication acceptance at *Teen Ink* or the university literary magazine. These happy moments are savored but then released as part of the transience of the moment, developments aligned with the third and fourth components of Right Effort. However, it's important that emotional zenith and writers' glory moments do not become causes of suffering, which will ensue if writers are unable to detach from peak experiences and move on to the next moment. I also aspire to have students better perceive writing emotions and, through observant acceptance, not be obliviously beholden to them. A non-evaluative acceptance of writing-related emotions counteracts the waste of mental energy that happens when writers deny those emotions. On top of neutrality, detachment, and ambivalence and those cultivated positive writing emotions, I want to introduce a replacement emotion, and it's this other emotion that serves as the predominant effect of a mindful writing class. Specifically, I want student writers to find a mild happiness, a Buddha's smile for writing, or a peacefulness that comes from abiding with the changes of the writing moment: it's the emotion of mindful metacognition.

Mind waves and mind weeds can be separately discussed or combined in a single activity, with the distinction made that mind waves more clearly occupy a specific present moment. Weeds last longer, so they're more obvious rhetorical factors in an ever-shifting present situation. It would be nearly impossible to recount a prior experience of mind waves, since their fleeting qualities make them unmemorable—just as a literal ocean wave replaces the next in our attention. In turn, this unremarkable quality can function as a writing asset, since students who inspect mind waves notice the constant impermanence and interconnection of waves. For example, as a way to underscore impermanence as a rhetorical factor in tandem with a second rhetorical factor of a mind wave, students observe their breathing and then register changes in body (pulse, breathing) or splashes of emotion during the few seconds after hearing of a new writing project, keeping a running tally. This "Observing the Waves" exercise can be altered for a specific phase in a writing process, such that students repeat this activity several times while working on a single project, noting the mind waves that arise with prewriting, drafting, rewriting, feedback, proofreading, editing, and turning the final version over to a reader or in conjunction with different genre. In another variation, mind waves are observed between contrastive acts of low- or high-stakes tasks: for instance, taking a minute to register mind waves right before a freewrite or momentwrite versus right before a last-stage draft. It's also important to note the state of being that happens immediately after a mind wave has moved away: to what extent can student writers return to a bare awareness after a wave? Moreover, working with mind waves complements efforts with mind weeds; examining a mind weed for its accompanying sensations helps writers detach from the weed and perceive it as part of a larger affective context. It diminishes the self-rhetorical power of a mind weed to be associated with less long-term emotional responses, and this highlights the fleeting nature of the weed, especially useful for emotional heavy hitters like writing perfectionism. Finally, a more retrospective activity in perceiving mind waves and mind weeds involves making a list of writing fears and hopes (general rather than task or moment specific) and privately freewriting about each item on the list. As students complete this freewrite, they simultaneously observe the waves and weeds that arise as they contemplate the affective wish list.

Several mindful writing activities pertaining to affect address learning to abide with challenging feelings about a specific writing task or general writing ability. In "The Fireplace," the intent is to stick with a heightened writing-related emotion (stress, fear, joy, thrill, etc.), one with such fervent self-pathos that it would normally completely coopt our consciousness, and make it the centerpiece of our attention. Instead of ignoring the emotion, if negative, or slipping into a fantastical storyline, if a positive emotion, writers focus their efforts on observing the feeling for its changes in imagery, physical reactions, and intensity. Much like a fire, any energetic writing-related emotion is in a state of flux, and this insight erodes any tendency to think of emotions as monolithic. That surge of dread that has mindlessly filled half a morning of trying to write now becomes an object of curiosity—its pea green nausea mixing with a fog of helplessness followed by changes in shape and size as it subtly turns over to another emotion like embarrassment or self-humor—or how the red heat inside a sensation of anger actually mingles with opaque and cooler bits because we're not 100% angry. The benefit to writers of this focused contemplation of affect is that, paradoxically, the troubling emotion invariably diminishes, fades, and altogether exits from the present rhetorical scene—a development that happens because of the writer's perceptions of the actual impermanence of the affective phenomenon. Noticing changes in our writing emotions is one step toward becoming more perceptive of the affective dimensions; it's an improvement from generalizing emotions or rendering them static. As Claudia explains, "I run through different emotions when I write: when I begin, I am anxious, and then halfway through I become bored with what I am writing, and at the end I get nervous because the piece didn't turn out to be what I hoped." Through a mindful detachment, student writers also reach a more nuanced, less polarizing view of their writing emotions, as Allison explains: "My anxiety about writing doesn't directly harm my writing, but it also doesn't help me either." Again, the constant change of the present rhetorical situation is a writer's resource, not burden. Impermanence modulates every rhetorical factor, including self-pathos and attitudinal constraints; the import of this situation is that nothing is fixed, locked into place, a permanent condition, you'll-have-to-live-with it, or inherent. It's likely that student writers will find it challenging to remain focused on a single affective formation—even bored by the effort—and suddenly that which has even seriously vexed the writer is not much of a bother.

What arises in student writers' consciousness when they are told their current writing abilities are already perfect, already Buddha, as they are right now, no need to change anything? Asking students to react to the idea of "already perfect" in a private freewrite attends to a prevailing affective experience among writers of perfectionism and debilitating standards conveyed through intrapersonal rhetoric. Students asked to contemplate their immediate perfection are initially skeptical but make connections between lowering one's writing standards and productivity; as Allison admitted, "if I could shut off my little editor-in-the-head, I could produce much more content and beef up my writing." Amanda saw the benefits of considering her perfection as a matter of intrapersonal voice: "If I could stop trying to fix myself as a writer, I would have a stronger voice, be less afraid of my audience, have more confidence, and turn in assignments without hesitation." By operating from the belief that students are "already okay as they are," a writing instructor counters the societal imperative for constant improvement and the cultural myth of the driven student, those "offspring of driven adults, aided and abetted by driven teachers" who fulfill what Geraldine DeLuca describes as the "scripted narrative that drives our country and is hard to interrupt, a deadly enacting of monkey mind" (34). First-year composition students release the emotions of perfectionism by selecting a part of the writing process or a structural component that has previously heightened tension and reflect on the mind waves and weeds that it generates. Starting a piece seems an excellent candidate for this intrapersonal work-out, and once students mindfully observe the various affective responses to this thorny phase, they'll likely be better able to find ideas for content or process.

The majority of first-year composition students carry around a stone backpack of self-evaluation, heavy with perfectionism, and in another assignment, "The Stone Backpack of Perfectionism," students complete an essay in which they describe lugging around an emotional backpack, exploring the "contents" of their backpack (preconceptions about writing, audience types and audience relationships, intrapersonal dialog), including documents, assignments, and writing materials that are complicit in this perfectionism. (I have also invited students to substitute perfectionism for another frequent and potent writing-related emotion of their choosing.) In a micro essay on perfectionism, Ruby possesses three backpacks: a "real backpack" made by Adidas, a "virtual backpack" containing emailed assignments, online learning platform notifications, and electronic files, and an "emotional backpack" made of boulders. The real backpack is chaotic: "I never throw anything away, so I also have an abundance of miscellaneous notes, assignments, or readings that I printed out—unnecessary memories" that make Ruby "recall a feeling—or maybe wondering what I received as a grade and whether the grade is on [the online learning platform]." To complicate matters, inside her virtual backpack lives a worm: "While I am typing an essay, it will eat my words—so I write something, as I'm writing, it disappears. Letters go away one by one—as if I'm tapping a backspace." Allison associates starting a new assignment with a stranger standing behind her holding her down by the backpack "with the force of one hundred men," so "[t]his give me the sense that I have nowhere to go." Although she changes in size in an Alice-in-Wonderland fashion as she tries to write, the backpack increases in heft as each choice she makes to proceed becomes a piece of quartz added to the bag, to the point where the heft is like "I'm carrying seventeen cinder blocks" (each representing a confusing assignment prompt). By the essay's end, working with the dream-like imagery has released Allison from her writing anxiety, and writing seems a more natural activity. Students are welcome to alter the material of the backpack (after all, it is in flux), make it more or less ethereal or invisible to other people, and show themselves occasionally free of the backpack, for instance, leaving it in another student's dorm room during a less inhibiting writing experience. The backpack can mutate into other types of metaphoric luggage or handbags to suggest that writing-related perfectionism need not be bound to school settings; however, a backpack is an easy symbol of student experience, and its placement on the back of a student is connotatively loaded.

To improve control of self-pathos, students work with a series of charged notions of increasing potency—from more minor occasions of writing discomfort to large-scale scenarios of panic. Advanced practice with writing-related self-pathos involves the purposeful introduction of an emotionally charged writing-related thought into their internal dialog and observing the resultant mental aftershock in a present rhetorical moment. Buddhist monks situate themselves in emotionally charged surroundings like nighttime burial grounds to hone their ability to remain non-reactive to events and to avoid filling the present moment with preconceptions (fearful ones). In the Bhayabherava Sutra, the Buddha is asked whether it's not an excessive distraction for a monk to meditate in jungles containing dangerous animals and other perils. The Buddha posits a scenario: "There are the specially auspicious nights of the fourteenth, the fifteenth, and the eighth of the fortnight. Now what if, on such nights as these, I were to dwell in such awe-inspiring, horrifying abodes as orchard shrines, woodland shrines, and tree shrines? Perhaps I might encounter that fear and dread" (Ñāṇamolí and Bodhi 104). The Buddha's advice is not to run away from the fearful situation but to ask, "'Why do I dwell always expecting fear and dread? What if I subdue that fear and dread while keeping the same posture that I am in when it comes upon me?'" (Ñāṇamolí and Bodhi 104). The goal is to observe with detachment and acceptance the various sensations of the charged formation without trying to alter or vanquish it. In an advanced practice, first-year composition students practice this balanced reaction to more extreme affective circumstances through informal reflection activities around particularly charged notions (failing a course is an obvious choice) or assignment scenarios.

An easily initiated practice in accepting writing affect is "Bowing to Difficulty," in which writers close a writing session by bowing (physically or mentally) to their desk, laptop, notebook, or other writing materials for the challenges presented during a session. It's a variant of a remark I once heard a scientist make during a radio interview about his process of returning over hundreds of days to his lab to perform an experiment without success: "Another day, another fail." Eventually, he obtained interesting results (and was interviewed on National Public Radio), but success wasn't guaranteed—a lack of control that in itself constitutes a potent form of difficulty for many writers. "Another day, another fail" is an equalizing statement, putting the same weight on "day" as on "fail." To work without expectation of outcome or product and without expectation of ease is a trademark skill of a mindful writer who is comfortable enough with uncertainty and detached enough to perceive variability. A writing moment or hour passes: it contains the failure of seemingly insufficient quantity or quality. It's a matter of indifference, since the basis of success is the ability to perceive the writing moment and not whether the writing moment yielded appreciated material. "[*M*]*astery* and *control* are deeply built into our model of writing," according to Peter Elbow, who felt that freewriting permits "passivity, an experience of nonstriving, unclenching, letting go," outlooks also possible through bowing to difficulty ("Toward" 131–132).

Mike Rose famously described unstuck students as people who were able to remain flexible around composing rules—with the exception of the one rule, to which they abided, that they would constantly keep trying, which I would argue is the very meaning of a writing discipline ("Rigid" 131–133). Right Discipline entails the return to a moment without adding anything extra in terms of evaluation to the experience, including resistance to obstacles. If I don't write a single word, so be it. If I write a ton, so be it.

Ultimately, the ability to mindfully write leads to the gray rock of ambivalence about writing or an equipoise that comes from not evaluating writing capacity: during times when we seem to possess the ability as well as times when we seem to lack the ability to progress. This strategy entails not adding to the present rhetorical moment—no emotional additives. We don't add the judgment that it's positive that we (or others) are able to produce writing or that it's negative that we (or others) are unable to write at that time. The act of writing is neither special nor commendable: it is simply an aspect that arose during a present moment, a motion in time as basic as breathing. This ambivalent productivity alters our regard of a writing occasion: a single phrase produced during a session carries the same weight as 500 words, which carries the same value as no words produced at all. However, student writers may believe that what they're writing is worthless or boring because some excitement or even trouble is not accompanying the writing activity. The trick with ambivalent productivity is to accept a writing experience relatively devoid of emotional nadir and zenith; the very ability to produce words, especially for worried students, may seem to warrant positive reaction. An interesting consequence of remaining unattached to writing affect is that less mental energy is expended in dealing with or trying to evoke writing-related emotions. Ideally, writing isn't draining, since to be exhausted by this way of writing would be as nonsensical as becoming tired from sitting in a chair under normal circumstances.

In "Exaggerate a Storyline," writers take their monkey minds into their own hands by purposefully extending a preconception about their writing ability in a mindful adaptation of a literacy narrative, that assignment routinely given in composition courses. Normally, a mindfulness practice discourages us from becoming so enchanted by a mental formation that we abandon present awareness to follow the formation into full-blown daydream. With this activity, students are instead encouraged to engage—albeit purposefully—a writing-related fantasy and make it the subject of their intrapersonal rhetoric for a stretch. A current assignment could serve as the trigger to the storyline, but its effects should (imaginatively) last longer in the rendered story than the time it takes to receive a final grade for that assignment. Students take the preconception to its fullest conclusions by developing a story in which there are plot developments (negative or positive), settings, other people, and consequences to the mental formation. What might come to pass in two days, a week, a month, a year, and in the long term in this preconception about this writing task? Bronwyn Williams makes a case for repurposing the ubiquitous literacy narrative assignment often given in first-year writing courses such that the assignment becomes an opportunity for students to unsettle the usual stories they tell themselves about their writing self-efficacy and construct an alternative writing identity (*Literacy* 41–42). Similarly, my exercise is a type of literacy narrative, except that it asks students to start from a dubious self-conception, and the fictional nature of the story automatically draws attention to the constructed nature of student representations of writing identity. Normally, a mind weed permitted to evolve into a full-blown preconception just registers its opinions with us and departs from the scene of consciousness, leaving us to deal with the fallout. In contrast, with this activity, writers take charge of their intra-personal tendencies and extend the mind weed beyond its normal time in the mind—pulling its images into a story that occupies time, adding images to the point where it's no longer an evaporating fantasy but a recorded piece of fiction.

5

THEIR ABILITY TO WRITE IS ALWAYS PRESENT

A Disciplinary Context for Mindfulness

Mindlessness is par for the course: it's pernicious in every human arena, so it should come as no surprise that mindlessness surfaces in how people teach writing to others. It's commonplace for people to mindlessly eat, mindlessly talk, mindlessly walk, mindlessly exercise, mindlessly drive cars, mindlessly clean, mindlessly pick up a cellphone, mindlessly sit, mindlessly gesture, mindlessly breathe, mindlessly think, mindlessly type, mindlessly turn pages, mindlessly check email, mindlessly cook, mindlessly parent, mindlessly produce, mindlessly buy, mindlessly dress, mindlessly build, mindlessly argue, mindlessly work, so it's to be expected that most of us sitting in a writing classroom mindlessly write and mindlessly teach. Although mindfulness for obvious reasons is a matter on my mind as I write this book, I can go an entire half day without consciously realizing the present. I am fully present while typing this sentence, but that wasn't the case during the previous three edits. Moreover, when I teach writing, I am merely one member of a larger system of mindless writing, as will be discussed in this chapter, because the standards for writing instruction at the national disciplinary level are for the most part systems of mindlessness. Along with the preceding list of activities, people mindlessly administer writing programs, developing mindless writing curricula, based on best practices in an academic discipline that preserves an approach to writing based on ignoring the present rhetorical moment. The difference is that mindlessly brushing one's teeth has little impact on others, but writing curricula and policies systematize mindlessness and organize the outlook and choices, minimally, of two million college students every academic year, extending a pattern that likely began in students' earlier education into the writing of the rest of their lives.

How can we strive to become mindful writing instructors inside what is quite possibly a mindless institution? After all, spiritual innovators throughout the ages and from a variety of religious traditions usually didn't linger inside problematic churches or religious systems but instead founded separate break-away institutions—the Buddha being no exception in his dealings with the Brahmins. At first glance, it may seem an impossibly seismic shift, a pairing of intractable opposites: all the future-cast, assessed endeavor that is twenty-first-century composition pedagogy and theory, on the one hand, and abiding with the moment through its shifts, uncertainties, defects, on the other. In addition to best practices ratified by national writing policy, even routine aspects of the composition classroom, such as assignment design and grading, might look incompatible with mindfulness. An expectation as basic as that we guide student writers to reach final drafts ready for grading might appear to be at the expense of the in-progress, the present-based, and the momentary. Are the generalizations inherent in learning outcomes, goals, and objectives at all compatible with the moment-specific, ever-fluctuating experience of mindfulness?

On the surface, national writing policy from major academic organizations such as the Conference of College Composition and Communication (CCCC) or the National Council of Teachers of English (NCTE) seems to comport with mindful learning. The policies draw attention to particular contexts and value pluralities, whether multiplicity of strategies, rhetorical components, points-of-view, purposes, or audiences. The emphasis on context and pluralities appears comparable to the continuous development of categories, receptivity to new information, and recognition of multiple perspectives that comprise mindful learning (Langer *Mindfulness* 37; 62). For instance, first-year composition learning outcomes touted in the Council of Writing Program Administrators' 2014 "Outcomes Statement" include the development of "facility in responding to a variety of situations and contexts calling for purposeful shifts in voice, tone, level of formality, design, medium, and/or structure" and the ability to "understand and use a variety of technologies to address a range of audiences." The policy recommends flexible instruction in writing processes so that students "adapt their composing processes to different contexts and occasions." CCCC's "Principles for the Postsecondary Teaching of Writing" tries to strike a balance between generalized and particular aspects of context, maintaining that composing means considering genres and the needs of writers as both collective and specific; for example, that student writers are taught to "study the expectations, values and norms associated with writing in specific contexts." NCTE's "Professional Knowledge for the Teaching of Writing" advocates helping student writers build a "repertory of routines, skills, strategies, and practices for generating, revising, and editing different kinds of text." This particular policy underscores the lifelong nature of learning to write, that students "develop and refine writing skills throughout their writing lives" and "do not accumulate process skills and strategies once and for all": a view in accord with Ellen Langer's opinion that instruction that pursues mastery shortchanges inventive thinking. The paramount way to reduce mindlessness in education is to teach students to notice an ever-shifting context and avoid static categories that can come from an overemphasis on memorization of the basics and rules (Langer and Moldoveanu "Construct" 3). The pursuit of newness is repeatedly mentioned in writing policy, which espouses students' ability "to use novel approaches for generating, investigating, and representing ideas" (Framework 4) or defines creativity as exploring material that is "new to them" or employing "methods that are new to them" (5). In addition to this handling of context, national writing policies appear to meet hallmarks of mindful learning, with several of the eight habits of mind endorsed by the Framework: openness or the "willingness to consider new ways of being"; flexibility or "the ability to adapt to situations, expectations, or demands"; and metacognition or "the ability to reflect on one's own thinking." The NCTE guidelines claim that writing should support "personal and spiritual growth" and that writing includes "non-public uses" listed as "self-organization, reflection, planning, and management of information." The crafting of policy statements, as efforts to summarize and advocate for the best pedagogical practices in the field, is undoubtedly a herculean task, but a closer look at the discipline's future-oriented rhetorics reveals that it actually promotes mindless learning.

In this chapter, I discuss the implications of a prevailing mindlessness in the institutional contexts of writing instruction, specifically how teaching practices with a future slant impede the potential of the present writing moment and create suffering for student writers and their instructors. Mindfulness pedagogy shifts focus from future textual productions and onto the real-time, fluctuating experiences and perceptions of writers, but it's an approach not evident in the discipline's pedagogic standards. To make this point, I examine national writing policies from the discipline's major organizations: the 2011 "Framework for Success in Postsecondary Writing," the 2014 "WPA Outcomes Statement for First-Year Composition," the 2015 CCCC "Principles for the Postsecondary Teaching of Writing," and the 2016 NCTE "Professional Knowledge for the Teaching of Writing." In these policies, mindlessness manifests in the downplaying of interiority and writer embodiment, the mishandling of the preverbal formlessness of consciousness, and the omission of the affective experience of student writers, not the least of which is the suffering caused by its own writing instruction. Of the problems inherent at the national policy level, the most consequential is the omission of the present moment for composing. This disciplinary mindlessness filters into writing instruction on a school-by-school basis, affecting the next level of curricular and policy documents at colleges and universities, with implications for around two million students each year and their instructors. My own department followed these documents when we revamped our first-year writing curriculum a few years ago, and I suspect hundreds if not thousands of curriculum documents in colleges and universities across the United States adopt the discipline's policies of mindless writing.

Yet it is possible to try for large-scale institutional changes that move a whole body of people toward mindfulness. The good news is that signs of mindfulness are already apparent in these national policies. In *A Mindful Nation*, Congressman Tim

Ryan, Democrat from Ohio, has proposed this revolution of the collective value system of the entire United States. Advocating for a governmentally supported mindfulness similar to the moon program of the Sputnik era, Ryan says:

> Right now the mindfulness revolution is an organic, grassroots movement. In order for that to infuse all aspects of our society, its research and implementation should have the helping hand of government. What we are talking about here is a growing personal and communal journey that can help us bring out our country's best. It's a way to increase our awareness of what's important to us, and what our priorities are both individually and as a nation.
>
> *(161)*

Like Christy Wenger, I spot an opening for mindfulness in national educational policies and a way to embark on conversations about system-wide mindfulness. Wenger sees an implied role for mindfulness in the "habits of mind" advocated in the "Framework for Success in Postsecondary Writing," especially in the document's pairing of rhetorical training with the development of awareness. Wenger concludes that "we might see the Framework as underscoring the importance of developed writerly awareness, or of approaching writing mindfully" (*Yoga* 104).

As another example, Robert Yagelski's ontological approach in *Writing as a Way of Being* is a significant step in the right direction. At its core, mindfulness pedagogy calls for a transformation in the discipline such that writing becomes a similar ontological act. Yagelski reorganizes writing instruction around the present moment, suggesting that the discipline replace its focus on student writers' texts with a focus on "writers writing," an ontological activity or way of being in the world. The temporality of this ontological approach is present-based; he repeatedly evokes the present with writing that "focuses on what happens *now* rather than what happens later," says writing is a way to "affirm and proclaim my being in the here and now," and points to how "the act of writing intensifies the writer's awareness of him or herself *at the moment of writing*" (107; 104; 112). Yagelski's ideal writing scenario is the National Writing Project, when an assembly writes together in the same vast conference meeting room. Yagelski reflects on how the "writing we did in those meetings was almost never intended to be revised or published or even ready by anyone other than the writer; the audience was usually ourselves at that moment. Sometimes we shared the writing, sometimes not. And once the meetings ended, the writing was usually disregarded" (xiv). Essentially, this scene of writing—collective, almost sangha-like—involves a shared group experience of writing that turns to the present moment and to each person's present experiences free of the restrictions that come from the future (expectations or evaluation). Yagelski wants to liberate the student masses who enroll in first-year compositions courses from assignments leading to "sanctioned texts that matter little to them or to us" by guiding them to this collective, low-stakes writing (171). As he declares, "This is the future of writing instruction I envision. 1,000 writers writing. Together" (165). Yagelski commendably adjusts our focus, yet provides few specifics, admitting, "I do not know how likely it is that we could realize this vision of 1,000 writers writing" (171). What I am saying in this chapter is that this vision of what might be possible for first-year writing instruction doesn't need to stop at massive sessions of publicly composed private writing. Approaches to the present rhetorical moment and to a present-based writing process of mindfulness pedagogy can actualize this goal on a larger institutional level.

It's my view that national writing policy is generally pointed in the right direction, but that the discipline hasn't developed a method to fully achieve its goals and, moreover, that in some regards policy actually interferes with mindful learning. It's clear from the predominant audience and discourse conventions in national writing policies that social views on composing that mishandle the writing self are deeply normalized in the field, in conjunction with non-present rhetorical practices and approaches to *kairos* that lack present awareness. These ideas about writing pervade rhetorical, genre, and process awareness, mainstay learning outcomes of first-year writing, and consequently mainstay assignments such as rhetorical and genre analysis. We need to teach writing students to notice rhetorical elements (like their intrapersonal rhetoric) on the backdrop of a present-based metacognition and to notice and accept the constant change of those elements. In this chapter, I discuss how mindlessness manifests in national writing policies through the neglect of present writing temporality, a mismanagement of the writing self, and underestimation of the fragmentary and nonverbal. Mindlessness in writing policy can be redressed through learning outcomes that emphasize process and rhetorical theory based on the present rhetorical moment. In many regards, composition scholars and teachers are on their way to being *bodhisattva* of writing—we are a group of professionals who have dedicated years of effort to helping others improve the paramount human activity of writing.

Overlooking Present Temporality

The most egregious problem lies in how the temporalities of writing are routinely framed in established writing pedagogy—a paucity of present moment awareness and instead, javelin tosses to a hypothetical future day in which a text is received by a reader. A writing context is never actualized if removed from its present context. Much of the language of the temporality of writing in national policy detours from the present. For instance, the 2015 CCCC "Position Statement" suggests that students should be shown how writing changes "over time," as in "Writing, like thinking, takes shape over time." If we were to change just the preposition to "in time," the sense that time is a requirement (as in writing takes time) with an outcome ("over" time implies an end goal) would be replaced by the notion of writing as an experience, one that unfolds in a series of moments in an open-ended way.

Making the same change in the statement that "[w]riting development takes place over time as students encounter different contexts, tasks, audiences, and purposes" in the "Framework for Success in Postsecondary Writing" would represent a step toward mindfulness. Interestingly, it's often little words such as "as," "in," "during," and "when" in these policies that are revealing of their authors' views on the time of writing. In NCTE's position statement, "when" is prominently used in an attempt, I would suggest, to hint at real-time writing activity. This document talks about "new thinking" which happens "when writers revise" and the "varieties of thinking people do when they compose, and what those types of thinking look like when they appear in writing." Subtle shifts, as with conceptual metaphor around composing, will do quick work toward moving instruction closer to mindfulness. If the wording could shift from the sense of "vary over time" to "vary in time," the present rhetorical moment and its fluctuations would be better highlighted.

A related mishandling of present writing time in national writing policies concerns the treatment of metacognition. On first inspection, what could be more seamlessly aligned with a mindfulness approach than the ability to observe one's thinking, essentially to notice one's passing intrapersonal rhetoric? As I mentioned earlier, however, this activity becomes separated from the present writing moment when it's cast as a matter of retrospection rather than in-progress, on-now, in real time. Instead of observing one's writing-related thinking when it stands in greatest proximity (within milliseconds) to the activity of writing, the advice in these documents is to help students reflect on process and rhetorical choices, on writing experiences, after those experiences have passed. The developers of the Framework suggest that metacognition, along with seven other identified habits of mind—curiosity, openness, engagement, creativity, persistence, responsibility, and flexibility—occur *as a result of* experiences with rhetoric, critical thinking, writing processes, discourse conventions, and multimodal composing environments rather than *during* those experiences. (The very notion of mental habits might be counterproductive from a mindfulness perspective, since the imperative to "reinforce the habits of mind" calls up mastery and drill-based approaches that Langer attributes to the automatic responses of mindless education.) As the Framework describes metacognition, it's the ability to "reflect on the texts that they have produced" and "connect choices they have made in texts" and to "use what they learn from reflections on one writing project to improve writing on subsequent projects"—the last point actually a deviation from the present that skips ahead to the future. Similarly, in the "WPA Outcomes Statement of First-Year Composition," metacognition is explained as the ability to "reflect on the development of composing practices and how those practices influence their work." In "Principles for Postsecondary Writing Instruction," a real-time setting for metacognition is slightly alluded to in the idea that instruction "includes explicit attention to interactions between metacognitive awareness and writing activity," though no further details are specified about this temporal interface, that shared moment.

In his study of metacognition, John H. Flavell drew a distinction between metacognitive knowledge found and applied in a particular moment and the use of a collection of metacognitive experiences in a new situation (Stewart et al. 1). The field should embrace a range of metacognitive experiences and include ones that occur in real time alongside more retrospective or accumulated ones.

In fact, Buddhist mindfulness is a better alternative to metacognition in writing instruction, and its adoption would fix many of the problems of institutional writing policy. It would put the brakes on future-oriented rhetorics and teaching practices, or at the very least counterbalance their effect of mindless learning. Buddhist mindfulness is a more sophisticated approach to metacognition because it happens *in real time* (not retrospective and not anticipatory). Because Buddhist metacognition is "live," it unfolds in time a series of ongoing developments and conducts attention to that stream of change. Subsequently, there's less mental room for trucked-in assumptions, because it's a fulltime job trying to keep up with the continuous procession of changes in state, outlook, and ideas of *anicca*. The next difference between Buddhist metacognition and that of writing policies is how the former adds the element of detachment. It's a clear instruction in letting go of whatever is noticed during metacognitive effort, which stays at the level of suggestion in the standard notions of flexibility and open-mindedness. To be in synch with impermanence causes fewer problems with ego, because the more we understand that entities are not static and discrete, the less likely we are to perceive *ourselves* in static and isolated ways. Combined with a willingness to let go and accept the comings and goings of the immediate moment, this metacognitive arrangement preempts writing adversity and leads to a state of grace for writers. Conversely, we recall that writing-related suffering results from letting ourselves become beguiled by mental formations, permitting our minds to cling to certain formations and experiences and reject others. Detachment is an essential cognitive activity in metacognition. The bonus of training in *sūnyatā* makes Buddhist meta-awareness an even more advanced form of metacognition than that evident in the field's curricular policies; with verbal emptiness, writers are given more help with what to do immediately after they perceive the contents of the present (return to bare contemplation). It's the crucial difference between telling writers to think about their thinking, which runs the risk of letting those mental formations dominate awareness (standard metacognition), and telling writers to stay focused on awareness itself (mindful metacognition). This added insurance makes it less likely that writers will slide back into habits of suffering.

In the policies of the discipline, writing is said to occupy time without occupying a moment at hand. The refrain in the policies that writing takes time commendably moves away from static notions of written pieces (that they can be accomplished in one sitting or that texts don't evolve over time) as well as from static notions of student writers (that a person's development in writing evolves over time). This readily accepted view in the discipline is usually matched with the notion of a recursive (and therefore complex) writing process. For example,

"Principles for Postsecondary Teaching of Writing" states that "Sound writing instruction recognizes writing processes as iterative and complex" and acknowledges that students "need time and feedback" and "writing is not produced in one sitting." The sense, however, is that time is a sort of commodity or resource (the discussion is about the use of time), and not much detail is provided on the nature of that resource or its specific qualities. Second, the recursivity of writing processes is framed in terms of big scheduling blocks (the phases of writing) with no attention, again, to the micro-changes and shifts of the moment. As one exception, NCTE's guidelines put instruction in "a repertory of routines, skills, strategies and practices for generating, revising, and editing different kinds of texts" with a metacognition that's more present focused, with student writers asking themselves questions that include "How do they plan the overall process, each section of their work, *and even the rest of the sentence they are writing right now*?" (emphasis added). In "Professional Knowledge for the Teaching of Writing," the long-term, extra-curricular development of writers is affirmed as teachers are asked "to consider what elements of their curriculum they could imagine students self-sponsoring outside school" with the goal that "writing has ample room to grow in individuals' lives," including that individuals "develop and refine writing skills throughout their writing lives." This expansive sense of the timing of writing has its uses (it's encouraging and pragmatic), but it's not a substitute for the real-time rhetorical resources of the moment.

The observation of time for the purposes of writing is cast as a single type of *kairos*, one that is more opportunistic and strategic than observational. For instance, in "Principles," it's explained that student writers "must understand how to take advantage of the opportunities with which they are presented and to address the constraints they encounter as they write. In practice, this means that writers learn to identify what is possible and not possible in diverse writing situations." Generally, discussions of *kairos* have pointed to one of two directions. The first view of *kairos* sets it up as an opportunistic gauging of the moment in order to size up audience and locate exigency. This strategy attempts to find the optimal means of persuasion (Kinneavy and Eskin 435). The sense of *kairos* as finding "the appropriateness of the discourse to the particular circumstances of the time, place, speaker, and audience involved" means that *kairos* becomes more "timely" than "time" (Kinneavy "Kairos: A Neglected" 84; Race; Rickert 75). In the second view, *kairos* shifts emphasis from strategizing to just noticing the momentary, passing qualities of present time for the purposes of rhetorical invention—a perspective found in the Sophistic traditions, where contingency and an ever-changing situational context were upheld over Platonic universal truths. What is needed (for at least part of instructional time) is a less evaluative, more detached perception of the fluctuation of the present rhetorical moment, with the intent of noticing changing content and outlook in order to reach a genuine multiplicity, flexibility, accurate metacognition. This second approach to *kairos* alters our notion of the rhetorical situation to include impermanence alongside exigence, audience, and constraints. After all, how can student writers avoid static notions about the act of writing and their own abilities—as well as about rhetoric and genre—if they are trained to overlook the fluctuations that surround them in the writing moment? And how could that omission *not* shortchange invention, since, as Langer says, to ignore flux is to reduce perceived possibilities, including the available means of persuasion? Writers without present awareness are acting with partial information.

Accordingly, activities, exercises, and assignments should balance both low and high stakes and should regularly ask students about the present moment. Some are freestanding studies of the present moment (breathing, posture, emotional formations) and some look at a present moment as it is affecting writing in tandem with writing (inventive shifts in attitude, passing intrapersonal rhetoric). This praxis infuses present awareness into descriptions/explanations of the rhetorical situation, so that it's not only about audience, exigence, and dealing with constraints, or about the various rhetorical moves, but the rhetorical situation has a setting, one in the Now. Revision is redefined as the ability to perceive change sans quality judgments of that changing material.

PROPOSED AMENDMENTS TO WRITING POLICY OUTCOMES

- Develop facility in shifting attention to the immediate, real-time rhetorical situation during writing, including setting, embodiment, and intrapersonal rhetoric.
- Develop flexibility by observing fluctuations in the writing moment with non-evaluative interest, acceptance, and detachment; employment of this practice during any writing moment at any stage in composing a written text and for a range of genre and assignments.
- Identify the ways in which student writers depart from present awareness and the rhetorical nature of future-bound ideation, including considerations of audience.

Genre instruction is a prime curricular area to revamp, since it would be easy to regress into mindlessness when dealing with socially approved forms. Genre signals to writers that they don't have to reinvent the wheel each time but can instead look for patterns in occasions and rely on pre-established forms (Dirk; Miller). We should take a page from Buddhist practitioners who know that any human activity, no matter how mundane or esoteric, sacred or profane, can be accomplished mindfully: setting a table, driving a car, walking, talking, engaging in an elaborate ritual. It's possible to follow a dress code or etiquette rule mindfully. The main question with genre is how genres function as social forms while also existing in a changing context; in actuality, each time a genre is used, the context is necessarily different and in flux. As a result, first-year students need to mindfully reflect on existent forms to search for subtle changes in both the genre and their experience of composing in it. Students should learn to see that genre is in flux by observing genre in different contexts, finding more than one genre for a type of content, finding new uses for a genre, and studying how a genre evolved over time. For instance, removing a genre from its usual context (a five-paragraph essay) and placing it in an unexpected one (a franchise restaurant) will, in addition to eliciting laughs (a sign of a productive discomfort), draw fresh attention to its features. Observing a genre in use demonstrates that genre as situated in particular rather than universal time. In a Langerian sense, students look for subtle modifications they can introduce as they work in the provided form. This activity could extend into assignments in which students repurpose a genre, such as through found writing, or even design their own genre for use in a reoccurring situation.

The Wrong No-Self Policy

The next haven of mindlessness in disciplinary-wide composition practices is the minimal amount of consideration given to a writing self: the bottom line is that we absolutely cannot have mindful metacognition without attention to the writing self as part of a rhetorical situation. When students write, they simultaneously generate and react to observations about the rhetorical situation through their intrapersonal rhetoric and through nonconceptual acts from the unconscious. At the same time, the embodied, physical selves of writing students fling up sensations and reactions to those sensations—all overlooked in current writing policy. This false type of no-self is connected to the broader omission of present temporality: to omit the self is to obscure a huge set of present-specific and contextual factors concerning the writer as a person—a prescription for mindlessness. Intrapersonal rhetoric, that constant verbal emission of the self, is not mentioned in writing policies. In fact, it's altogether missing from the "Developing Rhetorical Knowledge" section of the Framework, and NCTE's "Professional Knowledge" emphasizes the need for "deliberate insertions of opportunities for talk into the writing process," diverting student writers' intrapersonal talk during prewriting and composing to interpersonal talk between "trusted colleagues" and peers. In addition to the erasure of the primary text of intrapersonal rhetoric, this false sense of no-self is detrimental to the metacognitive skills the field seeks to teach through rhetorical and process awareness.

As an organization, we respect a type of metacognition that is oddly scrubbed of the observing self as well as dislocated from present circumstances. Warning that "teaching the *intellectual tools* of writing and rhetoric without accompanying *tools of awareness* makes writing instruction incomplete at best and dangerous at worst,"

Paula Mathieu describes how she remedies this situation by asking students to notice their "inner rhetoric and the ways to revise the voices in our heads" (17). Recently, Ellen C. Carillo argued that mindfulness adds the student self to the metacognitive endeavor of reading in first-year composition. According to Carillo, "the concept of mindfulness highlights not just the task that one does 'mindfully,' but the individual, the reader, who is learning *to be* mindful. Mindfulness, unlike metacognition, is a way of being" (118). The writing self is integral to the writing context: to abrogate the writing self is to propagate mindlessness by substituting all sorts of predetermined notions for the particularities of now. This erasure of the writing self is a consequence of persistent binaries between self and other, interior and exterior in the discipline, ones which a Buddhist sense of no-self could redress. At best, the Framework begins to suggest an ontological approach to writing instruction in how it explains the second habit of mind, openness, as involving the consideration of "new ways of being." As it stands, however, national writing policy mostly operates with a mistaken sense of no-self.

How bizarre it is that in so many conversations about composing pedagogy, the writing self in its myriad of affective, cognitive, and embodied aspects is disregarded, leaving a veritable silhouette of where the person has been unscrewed from the scene of writing! It would be absurd to claim that the way a dancer holds a wrist or positions a foot is nonconsequential to the dance or that a painter's brushstroke style, caused by the angle of hand or pressure applied, was immaterial to the still-life. It would be absurd to remove the physical dancer from the dance, a genre that occurs in real time; the challenge is to give the same potency of presence to genres in which the creator is separated in space and time from audience.

On a typical writing day, around 2:15 or 2:30 pm, I experience a host of physical sensations that are directly the result of my writing experience—my neck and shoulders are stiff from looking down at the laptop screen or at research materials, and I feel a dip in my blood sugars due to my tendency to carb up during a writing lunch, compounded by an espresso that's ebbed out. I feel the blood pull in my feet and crossed ankles from the hours of sitting at the desk. I am still digesting my lunch—there's a certain heaviness—plus a craving to go for a run to jumpstart circulation and metabolism. At the same time, aware of the imminent arrival of the 3 pm school bus and a door-bell-ringing middle-school daughter, I feel a campaign of adrenalin as I try to rush my ideas on to the page in the remaining time. All these physical factors, the consequences of the activity of prolonged writing, are invariably part of an individual writer's experience and, more importantly, exert influence on my composing process as much as a deadline or word count (arguably "external" rhetorical constraints). They'll affect the shape of my sentences, my ability to concentrate and make logical and intuitive connections, my receptivity to ideas. They'll affect what you're reading right now. I might feel more worn-down than I would have if I had opted for a salad lunch because of the dip in blood sugar, which might in turn steer how I perceive exigence, short and long term.

In the various policy statements of the discipline, not once is a physical sensation of writing or a physicality of a student writer mentioned, and neither are student writers' affective responses to the need to write. This peculiar absence of writing selves is occurring in policy despite the significant work by scholars including Kristie Fleckenstein, Abby Knoblauch, and Christy I. Wenger on embodied dimensions of writing. Regrettably, national composition policy is "making the body irrelevant to the words the body produces," preferring instead to focus on a distant, future-based, and arguably hypothetical audience (Yagelski 45). The student writers who receive the focus of these policy documents are so out-of-focus as to be disembodied. This situation is happening although, as Joel Wilson reminds us, "writing is, at its core, a bodily function—sans hand, what words would be inscribed on a page or typed into a keyboard?" (174). As Wilson describes it, formulaic instruction geared toward state assessments at the high-school level has "ossified" students' experiences of writing, denying them their writing rituals and eliminating somatic dimensions of writing while also "contribut[ing] to a sort of physical trepidation associated with the composing process" (174). Students are left with writing stress and other side effects from school assignments without the resources of embodiment and present awareness which would alleviate such unpleasantness.

Moreover, it's not just a matter of not including student writers' breathing or embodied responses; the denial of the self means that individuality and interiority seem altogether gone in these documents. Body is gone—but so is affective experience around writing—and so, for the most part, is any sense of interiority. The "WPA Outcomes Statement of First-Year Composition," which claims that it is aligned with the practices of the Framework, says that "a large body of research demonstrat[es] that the process of learning to write in any medium is complex: it is both individual and social and demands continued practice and informed guidance." However, its quartet of outcomes—rhetorical knowledge, critical reading/thinking/writing, process, and conventions—are devoid of the writing individual. A process outcome for first-year composition courses is to "experience the collaborative and social aspects of writing" and for rhetorical knowledge to "write for different audiences, purposes and contexts," followed by "write for real audiences and purposes, and analyze a writer's choices in light of those audiences and purposes." In "Principles for the Postsecondary Teaching of Writing," the student writer seems to be given equal standing with readers inside a rhetorical situation: "writing is 'rhetorical' means that writing is always shaped by a *combination of the purposes and expectations of writers and readers* and the uses that writing serves in specific contexts" (emphasis added). I take this to mean that a rhetorical situation will include factors that are generated by and of use to the writer—such as exigence and expectation. The remainder of "Principles" gives no consideration of the self, including in the document's coverage of "real audiences"; writing is framed as a social act involving at least one reader, with the suggestion that the student doesn't function as a reader to her own text. We make true ghost writers of our students, turning them into apparitional selves.

In discussing the flexibility needed to address a rhetorical situation, it's entirely a matter of an interpersonal rather than intrapersonal situation. Altogether omitting the constraints generated by the writer's mind in the rhetorical moment, the CCCC document states: "[Students] should be able to pursue their purposes by consciously adapting their writing both to the contexts in which it will be read and to the expectations, knowledge, experiences, values, and beliefs of their readers." On the other hand, NCTE's "Professional Knowledge for the Teaching of Writing" mentions internal engagement for the purposes of writing and includes the intrapersonal and the self as traditional rhetorical factors of audience and purpose: "Writing with certain purposes in mind, the writer focuses attention on what the audience is thinking or believing; *other times, the writer focuses more on the information she or he is organizing, or on her or his own emergent thoughts and feelings*" (emphasis added). The document suggests that writing can be used to "support personal and spiritual growth" and that excellent instruction in writing supports a "range of non-public uses of writing for self-organization, reflection, planning, and management of information." No wonder so many people feel deracinated, uprooted, abstract, not like themselves when they try to write. Student writers are trained to take their eyes away from the immediate writing circumstances and look ahead to a day in which their text is not only received by a reader but already written—a mindlessness normalized by instructions in audience, determining audience expectations, finding authentic audience, writing for a range of audiences.

In some composition circles, to talk about the writing self is to risk running afoul of the prevalent ideology of writing instruction and being labeled retrograde, naively or destructively Romantic, expressivist, solipsistic, capitalistic, asocial, an accomplice or guinea pig to hegemony, or hubristic. The very word "self" is enclosed in a mental electric fence. Privacy and solitude are no longer enjoyed as a working condition but instead become conflated with privilege, castigated as private property/gated community/garret. To talk of solitude for writing is to unfairly establish a prerequisite for becoming a writer, one that supposedly excludes anyone who needs to write at the kitchen counter between piles of their children's homework or on their car's steering wheel at lunch break. For example, Karen Burke LeFevre characterizes the individual writer as a Platonic invention "compatible with certain assumptions characteristic to Western capitalistic societies," in which the "writer is portrayed as an individual who is supremely self-reliant" (15; 25). In LeFevre's treatment, this situation leads to an unflattering portrait of writers as self-absorbed, "isolated, competitive, and aggressive" (26). With Linda Brodkey's construct of the scene of writing, a troubled image inherited from literary modernism, a writer is socially isolated (in a proverbial or literal garret), confined by their writing activity such that "writing costs writers their lives" (398). A writer, disempowered and trapped, is reduced to the role of transcriber, becoming an amanuensis rather than the generator of language: "the writer is an unwilling captive of language, which writes itself through the writer"

(398). The second crisis resulting from this scene of writing, according to Brodkey, is that writing becomes static, a "picture postcard of writing" that "encourages the reification of one moment in writing *as writing,* by excluding all other moments" (399; 400). (I'm not sure how a writing study can be both dire imprisonment and idyllic vacation spot.) In a sci-fi fashion, a writer's decision to seclude herself in a room to write lets her stop time: "The closed shutters of the garret, the drawn drapes of the study, or the walls of books lining the library all effectively remove the writer from time as well as space" (404). At the same time, in describing her own writing session, she emphasizes the temporal and the fleeting (emphasis added): "I am struck by how *transient* are the images of myself as a writer ... I can catch *a fleeting glimpse* of myself *as I move* toward the phone to call a friend about an idea that troubles me" (396). Through a mindfulness approach to composing, I would counter, each scene of writing is exactly this transient and fluid, if a writer is able to remain aware of the present rhetorical situation. Troubling thoughts such as those of loneliness are themselves mental formations arising through intrapersonal rhetoric in the writing moment.

At the same time, composition theorists who dismiss the solo performances of writers in favor of a social view frequently resort to the imagination to explain what exactly is social about composing. In other words, to make writing in the moment a social act, such that student writers are seemingly in contact and interactive with others, theorists have needed to populate the writing moment with imaginary beings. On a single page, Brodkey uses "imagine" and its variants four times: she upholds a writer who has the "ability to imagine himself in the company of others even as he sits alone writing," to "imagine the materiality of language itself," "to imagine oneself as a member of a community of writers," and to "experience, and imagine, writing as a social as well as cognitive act ... as a form of resistance" (414). Karen Burke LeFevre fills a room with imaginary friends when she claims "[o]ne invents in part because of others, because one thinks fruitfully in the company of a great many others, who are both possible and real" (93). Even Robert Yagelski, in explaining his ontological experience of writing, deviates from the literalness of the present writing moment as he explains "that I am intensely in the here and now while I write, and at the same time I am also connected to something larger that is not here and now," mentioning his connections with editors, reviewers, fellow composition professionals, earlier scholars, an outreach that extends to "the history of this field and of writing itself ... not to mention my own past" (103–104). From the vantage of mindfulness, the problem with these imaginative leaps is that they acerbate student writers' inclination to populate the act of writing with hypothetical, future-based readers constructed on assumptions carried in on their intrapersonal talk. This tendency toward an illusory solution to the literal non-presence of audience is evident in national writing policies.

In NCTE's guidelines, the fakeness of future audience is sidestepped through a sort of substitute audience. These individuals are more proximate and take the form of "other writers, friends, members of a given community" who are nearby "during the process of composing." Accordingly, writing "happens in the midst of a web of relationships" including ones with readers, but it's telling that those "particular people [who] surround the writer," the ones who might actually share a similar physical space with the student, receive much more agency. These individuals are not just "a definite idea" like the reader of the student's piece but are people who "may know what the writer is doing and be indirectly involved in it." Similarly, in CCCC's position paper, the notion that writing is almost invariably a social act is supported by the types of interactions a student writer might have in collaboration over a document "as they compose" as much as those with future readers. It seems that whenever we walk away from particular rhetorical moments to discuss the impact of others on writing, we resort to mirages—leaving the actual, the discussion turns fictional—or call upon people who are closer in time and space to the student writer to compensate for the absence of those future people. A more realistic perspective on the writing moment would bear more resemblance to Alice Brand's account: "When we are poised in a reflexive stance with our ideas, we are a community of one. When we sit at our desk, when we write at our computers, we are in the end alone ... only one person at a time ultimately holds the pencil" ("Social Cognition" 402). Incorporated in national writing policy, a mindfulness perspective would maintain that such a sense of connectedness is fundamentally a product of intrapersonal rhetoric and mental formations arising during the moment of writing.

To avoid the decontextualization of the moment that comes from depersonalizing it, we should seek a more literal sense of what happens when someone writes and resist the metaphoric. Not what we guess is happening or what we think ought to happen when people write, but what is actually happening in the real time of writing, and silence, literal, physical silence, might be a better conceptualization of writing. The writer's lips (unless a mouth breather) are closed, the tongue mostly dormant, the facial movements and other bodily movements are not those of a speaker. The actual sounds produced (typing, pen or pencil moving, pages turning) are not those of speaking. I haven't literally talked to anyone in the past ninety-four minutes, though I've heard the words I've been typing, the words and ideas that didn't reach the screen but stayed in my intrapersonal text, as well as the sounds of the response I sent in an email and the anticipation of talk if one of my daughters, moving around upstairs, decides to open my study door. I've changed position in my office chair, pulled a strand of hair behind an ear, typed, moved books and pages around, put a pen cap in my mouth, changed music, finished a coffee, and taken off a bangle and ring, but I have not engaged in the physical activity of speaking aloud. If another person happened to be hiding in my study without a clear view of my laptop screen, he or she would have no access to any of the words I've "said" in the last hour and a half.

The Buddhist construct of no-self offers insight into the interiority/exteriority dualism that may be prohibiting a realistic self in writing curricula and policies. A few writing theorists have addressed this binary—Patricia Bizzell's attempt to complement inner-directed with outer-directed theory and Thomas Kent's paralogic, notably—but none are as efficacious as a Buddhist approach to self. Kent's paralogic attempts to reconcile this dualism but ultimately bolsters dualism: he proposes an externalist view in which "no split exists between an inner and outer world," but as the name might suggest, an externalist view resigns the interior to a secondary position, remarking that "internal mental states derive from communicative interaction; communicative interaction does not derive from our internal mental states" (104; 117). So far, the net result is just more dualistic thinking. Meanwhile, in Buddhist theory, the very question of whether there exists a self who writes would be moot, since probably indicative of a mindless outlook. As George Kalamaras puts it, "unlike Western rationalism, meditative consciousness is an experience of unity in which paradoxes such as self/other, inner/outer, the seer/seen, personal/cultural, as well as other seeming contradictions, are resolved and the seer and the seen become one" ("Meditative" 24). According to Buddhism, we are empty because we are empty of independent existence and because our selves are comprised of our interconnection with other entities (Lopez 29). The notion of a self, on the other hand, presumes the possession of separateness that Buddhism maintains is always an illusion. As Jennifer McMahon Railey describes the dissolution of self, "when one analyzes one's condition honestly, one cannot find a self" (126). Railey points to the impermanence of the human body, in which "the substances which combine to form [it] are continually being replaced so that one's physical composition is not the same from one day to the next" and suggests that our thoughts are "constantly, and often radically, modified to the extent that sometimes one cannot believe one remains the same person" (126). In mindful composing pedagogy, the focus is on rising and falling intrapersonal rhetoric without labeling that phenomenon "self," "individual," or "property of."

Specifically, it is because of the intertextual constitution of a writer's intrapersonal rhetoric, an interior discussion made of the comings and goings of various phrases, ideas, images, voices, that no rigid boundary stands between "writer" and "other." Actually, a mindful theory of composition is not incompatible with the theory delineated by Karen Burke LeFevre and other social constructivists: the writer is affected by social norms; the writer uses socially established language; the writer contributes to a body of knowledge shared by society; writers think of other people as they invent; a writer is influenced by institutions which are social collectives; and the reception of a writer's work is affected by social context (33–35). LeFevre also wants to undo "unhelpful oppositions" of individual and social, seeking their interdependency, in which "[a] change in the individual influences social dimensions, which in turn influence the individual ... it is impossible to say which is first, or which predominates" (37). I applaud the increased fluidity of this bi-directional approach but would note the persistence of a "bi" or dualism: better still is the Buddhist no-self approach in which attention, yes, is given to a person's arising mental phenomena without making the self a falsely separate entity.

Replacing the notion of a writing self as almost a silhouette in complete relief from the rest of its context with the idea of a writer *as context* might further disperse the contents of each perceived writing moment—attitudinal factors, sensory perceptions, body awareness, preconceptions, intrapersonal rhetoric—not binding it to ego. This view is in accord with Suzuki's ideas about small and big mind, in which "[i]f your mind is related to something outside itself, that mind is a small mind, a limited mind. If your mind is not related to anything else, then there is no dualistic understanding in the activity of your mind" (35). Ego dissolves into perceived rhetorical factors, which in themselves evaporate and are replaced by other factors, leading to a constant attention to the rhetorical context. As Suzuki says, "Big mind experiences everything within itself" (35). These moments of writing come down to an intertextually complex inner language manifesting on the vast surface of awareness with no stipulations concerning who that language "belongs to" or if there's a fixed observer-writer.

A related issue of the writing self is the way in which new mental formations that arise during a writing moment are framed in certain composition theory. Frequently, a belief in the capacity of writers to produce new material has been maligned as a sort of naïve notion of originality, talent, indulgent expressivism. A prime example would be Donald Bartholomae's "invention" of the university in which he deflates expectations of original undergraduate work with what he thinks is the more reasonable expectation that they "invent" their role in academia by mimicking its discourse; their legitimate work lies in managing their responses as writing students to conventions. As he says, "Leading students to believe that they are responsible for something new or original, unless they understand what those words mean with regard to writing, is a dangerous and counterproductive practice" (632). In contrast, from a Buddhist perspective, a new moment of writing is not necessarily an interesting, ingenious, exciting, or admirable occasion. It can be downright banal, as unremarkable as dryer lint, easily relinquished. Each observed moment is an emanation of newness. This newness isn't a quality to be evaluated or sought; instead, it's a matter-of-fact temporal occurrence. This constant change doesn't warrant an added evaluation of "originality" because such thinking risks attachment, clinging, wanting to "keep" an idea or moment, and writing suffering may ensue.

From a mindfulness perspective, a student writer stands before a parade of new moments not sorted into their relevance or significance, such that a new sensation in the typing hand is just as new as a fresh thought on a working claim, a fragment is just as much part of the inventive moment as a paragraph. When Ellen Langer explained the new perceptions possible in mindful learning, she meant that learners look for gradations in an experience, not ranking those experiences, and that newness also came from considering each moment as another context to investigate (*Mindfulness* 11). Donald C. Jones' views on discovery offer a refreshing view of newness without any imposition of value. Jones says that the "emphasis on surprise and discovery [found in process proponents like Murray and Elbow]

PROPOSED AMENDMENTS TO WRITING POLICY OUTCOMES

- Draw students' attention to the physical aspects of writing in order to heighten awareness of the rhetorical moment for the purposes of metacognition and invention.
- Include student writers' affective responses to writing occasions as rhetorical factors in any composing situation.
- Develop student writers' awareness of intrapersonal rhetoric throughout a range of writing activities, both low and high stakes.
- Identify and analyze occasions of self-pathos, including preconceptions about writing ability and writing task, as they arise and examine them for their affective impact on the rhetorical situation and writing process.
- Develop calm outlook for writing through acceptance of passing developments of a writing moment and detached observation of arising emotions and images related to needing to write.

does not mean that students are conceived as the autonomous authors of their own knowledge" but that they are "active participants in this method of inquiry" (89). Invention results from engagement with what Jones calls experience or the individual stream of consciousness, referencing Dewey and William James (88). Writers focus directly on the flow of experience without trying to predict outcome, avoiding preconceptions in favor of immediate writing experience (89). In this way, constantly arising material isn't the pollution of the ego but, rather, the outcome of a watched moment.

Valuing Formlessness

To help students develop a more accurate perception of the present moment for the purposes of writing, first-year composition curricula need to actively encourage formlessness and the fragmentary, the nonverbal and the partial, as part of prewriting and mindful invention that continues into seemingly later stages such as revision and editing. This shift alters how we treat the less valued components of discursivity—including nonwriting—usually neglected by future-based rhetorics and teaching practices. Formlessness and fragments configure a present-based awareness and mitigate product orientation. To optimally succeed with writing, one's primary allegiance must be to the present moment, not to the text, to the now of process and not the future product, and this requires remaining fully attentive to the fragmentary, continuously mutating and migrating performances of now.

When we look closely at a moment, what we invariably see happens within the dimensions of a single breath or an inhalation and three-quarters to all of an exhalation: details from our immediate setting, a phrase or two, maybe the tail-end of an image, and maybe nonverbal sensory experience. Silence. How does that fragmentary scene of the moment correlate with the writing taught in our courses, with full sentences, let alone a complete essay? Why pursue such a foreshortened view, since wouldn't it be like putting on blinders, limiting our forward momentum as writers? The reason for adopting a view toward the writing moment is precisely because the moment is reliably discursive, and that's more than can be said for staring off into the future, waiting for writing to happen. Resistance to a moment-based focus in this way is similar to resistance to freewriting, the way some students vent, "Why would I want to waste my time producing material that won't resemble the final version I need to produce?" The answer is that the moment, like freewriting, is constantly and reliably generative, and that the moment is constantly and reliably generative *because* it is fleeting and fragmentary. Second, each writing moment is restorative and calming, especially important for students who operate under years of built-up stress, urgency, and the pressure to demonstrate understanding, perform ability. Third, a view toward the momentary changes the timing of writing and increases writers' flexibility as they make use of the moment, any moment, to write—without needing a special type of moment, interruptions are fine.

Accepting the fragmentary is important to a mindful writing curriculum because it represents tolerance of the fluctuations of the present moment and perception of intrapersonal arisings. The dimensions of any present rhetorical situation are dictated by the average human attention span, which typically can focus on the present for a few seconds. A full-blown essay never occupies the present moment: rather, a word, phrase, or in some cases a sentence are the only content of a moment. Prewriting needs to explicitly embrace the fragmentary as an indication that writers' intrapersonal rhetoric is actually observed and treated with non-evaluative acceptance. Arguing that composition studies has overly emphasized complete, intact texts through assignments and model readings, James Seitz advocates for more parataxis (fragmentary, elliptical) in student writing and less hypotaxis (stability, convention) (816; 818). Seitz's views on the fragment—that it possesses the "rhetorical and aesthetic power generated by disruption, discontinuity and disorder, as writers enter roles whose constraints demand unconventional perspectives and representations"—comports with a mindful pedagogy (823). With the fragment, static and solid-state ideas are replaced with attention to flux and interconnection (823). The paradox is how the fragmentary or partial suggests interconnection, more so than a completed whole text. A fragment is not freestanding or autonomous. It does not presume to be complete unto itself but is, rather, one element in a complex, ever-changing verbal emptiness.

A view favorable to the fragmentary can positively impact on our sense of the timing of the writing process and specifically what it means to begin to write.

A moment-based invention strategy, to refer to the advice of the musician John Cage, means we "Begin anywhere." What this means is that the inventive moment isn't specially cordoned off with preconceptions of what student writers and their teachers think constitutes a good arena for starting to work. It also means less judging of the present rhetorical moment, since any moment is suitable for a "beginning." If we operate through the other lens, turning to specially demarcated moments of starting, we are making the mistake of predetermining the content we expect to emerge out of that moment. On the other hand, a mindful approach to invention sees any rhetorical moment as viable, displacing our usual effort (of selection, screening, evaluation, sorting) into another effort (of observing, accepting, detaching). Again, any gesture that predetermines context preempts possibility, reducing the variations that can be perceived. Likewise, inventive revision means returning to verbal emptiness, to activities of prewriting, and to the fragmentary even when a writer possesses more fully formed work on the screen. Changing the temporal sequence of a written piece in order to reveal the actual moment-by-moment construction, those ant tracks of real time, hidden inside a polished read-top-to-bottom-left-to-right document, can help dissolve students' assumptions about the timing of types of writing acts. Purely hypothetically, the literal final act in writing a book on mindfulness might very well *not* be the last phrase on the last page or the final citation entry but some sort of prewriting, buried in the manuscript's middle, a return to verbal emptiness at the last hour on the 705th day of the project. Writing curricula would step away from the assignment prompt/discover & draft/rewrite/finish rinse and repeat cycle in favor of sequences with less definitive closure and more intermittent beginnings.

We would want to honor the nonverbal in student writers' experiences to diminish process/product and ability/non-ability binaries. By acknowledging and investigating nonverbal experiences in acts of composing, instructors and student writers let go of a preference for words and production, a preference that's apparent even in process pedagogy. The old chestnut "process" is a conceptual metaphor weighted toward an outcome: "it's a process" or "it takes a process" implies a progression that culminates in a sought-after end. Process pedagogies pursue organized end products as a result of their explorations of invention (Seitz 820). In this view, a process appears acceptable as long as it results in a product, and in writing policy, it's assumed that a composing process results in *something*; as the WPA "Outcomes Statement" innocuously puts it, "Develop a writing project through multiple drafts." It's expected that instructors guide students toward writing completed pieces at the expense of the unfinished, the in-progress, and the momentary. From a mindfulness perspective, such dualistic thinking, complicit with evaluative thinking and clinging, will invariably incur suffering. In this foreclosure of the writing moment, even if we practice mindful writing, if our intent is skewed toward an outcome, something to keep, we are shortchanging the inventive possibilities of the now. Instead, writing policy should speak to the importance of training student writers to dwell longer in verbal emptiness and assist that learning by prolonging prewriting activities in a course. As Don Murray pointed out, instructors often annex the openness of prewriting because they feel uncomfortable waiting for student first drafts to emerge; as a result, teachers make the mistake of replacing the uncertainty of inventive moments with assignments that specify topic and genre ("Writing" 3–4). This conundrum is well put in the NCTE policy: "Every teacher has to resolve a tension between writing as generating and shaping ideas and writing as a final product." To that end, assignments in first-year composition should include ones that distance themselves as much as possible from the forms of textual products, such as disposable writing, sustained prewriting, and no-writing. The latter involves purposefully not recording any passing intrapersonal thought and is probably the ultimate practice in non-attachment.

In the policies of the discipline, prewriting needs to better incorporate the preverbal or those moments in which a student turns to formlessness to activate his intrapersonal voice for the purposes of writing, as well as those moments of nonwriting to maintain a nondualistic approach to invention. With the preverbal, no language toward the designated effort has been produced—not notes, a freewrite, a brainstorm, or an outline. The preverbal entails the contemplation of emptiness before language rushes in; the preverbal condition is expansive, with no predeterminations made about content, genre, or the address of constraints. As a largely non-discursive cognitive terrain, the preverbal is an opportunity to explore important factors in the rhetorical moment, such as writing embodiment and affective responses. Nonwriting differs from the preverbal in that it requires the wholesale acceptance of no-product in an attempt to maintain bare awareness. Unlike prewriting, with a prefix of "pre" that suggests knowledge (or hope) that the next moment will generate writing, nonwriting (or no-writing) altogether drops that expectation (or hope) and abides in the open moment. Furthermore, the preverbal and nonwriting should not be confined in curricular policies to the stage prior to a draft, but should function recursively, such that students return again and again to formlessness throughout the experience of writing to find content, gain insight into their affective responses to the task, and reengage mindfully with the moment.

Instruction in the preverbal side of prewriting is crucial with undergraduate writers to prevent the misapprehension of nonwriting and the preverbal moment as a writing block. Otherwise, students might misconstrue the preverbal as a malformation in their writing ability, possibly as reason to give up altogether, instead of as a door to the rhetorical moment. Many student writers are trained to understand writing as a process in which a text transitions from formlessness to ever-increasing form with stability of structure. While it's possible for a piece of writing to linger in a more formless, expressive stage, as James Britton pointed out, most assignments guide and goad toward a revised text. If the impulse for polished writing overwhelms earlier exploratory moments, what results is a false emphasis on outcome that can lead to problems in composing. Instructors, too, will need guidance in accepting formlessness from their students, because the "teacher not only has to face blank papers but blank students worried by their blankness, and a blank curriculum which worries the teacher's supervisors" (Murray "Writing" 18). Because formlessness cannot be directly assessed, because it is invisible to the teacher and outside his or her control, formlessness may seem anti-curricular. However, these "non-productive moments"—or really "non" overall—offer student writers crucial experiences with verbal emptiness and interconnection. Instructors' uncertainty about verbal emptiness is alleviated when student writers compose process notes describing their efforts with this side of prewriting.

In actuality, form and formlessness are in interplay, because bolstering formlessness in writing curricula paradoxically strengthens form. Wonderfully, the more formlessness is attended to, accepted, and nurtured, the more likely it is that form—or writing material—will make an appearance. Again, this emergence occurs because of two factors: the inherent discursive nature of the human mind and the nature of impermanence, which necessitates that a phenomenon (such as nonwriting) will invariably be replaced by another phenomenon (starting to write). A place should be saved for formlessness at any stage in the chronology of writing a piece, and formlessness complements even the most "formed" writing assignments in first-year composition curricula such as those in a genre. "Principles for the Postsecondary Teaching of Writing" does a laudable job of pointing to the impermanence of genre—to the importance of teaching student writers to see that "many genres change over time." It should be pointed out that constant change of form happens through the return to formlessness, something which happens *as* student writers compose those genre.

Routine aspects of the composition classroom, like assignment design and grading, might seem to present instructors with roadblocks to mindfulness that would preclude the fragmentary or the fleeting. How can we reconcile the need to respond, evaluate, and assess student writing with the non-evaluative open acceptance a mindfulness-composing pedagogy seeks to instill? In terms of the partial and fleeting, assignments can be modified such that they give much more instructional time to helping students observe prewriting, perhaps even making prewriting the major part of an assignment. Kim Brian Lovejoy's "self-selected writing" assignments are a possible model in which students frequently write on topics of their own choosing in brief sessions and then decide which pieces to further develop or share; as many as thirty pieces are written over an eight-week period of time, and only four or five make it into a portfolio (85). Similarly, in a self-sponsored invention in a mindful writing class, students decide how long to remain in the nonverbal or preverbal before the moment is right to begin drafting. Along the way, they complete frequent process notes describing their encounters with the present rhetorical moment (its intrapersonal rhetoric, embodied and affective experiences, fluctuations) and the reasons for their necessary delays or move toward drafting. Another mindful assignment design asks students to make radical changes to an early draft over multiple occasions, such that completely different drafts result rather than a production line toward a unified, polished product. For the assessment of mindful writing, in broad strokes, in addition to supplying first-year students with experiences of complete non-feedback and non-evaluation, the grading system of the course recognizes students' attempt at developing mindfulness as much as (and perhaps more than) evaluating the quality of their written work. Contract grading and Asao B. Inoue's innovative method of assessing students on the basis of the "quantity of time and effort put into a project or an activity" rather than the quality of writing are methods to change the emphasis of evaluation to one more present friendly (Inoue 73). Similarly, students can be evaluated on the quantity of their attempts at mindfulness (how regular their practice); the length of their attempts (how long they practice); and the range of mindful writing techniques they implement (yoga for hands, freewriting, momentwriting, mindful breathing, Peter Elbow's Open-Ended method, Sondra Perl's felt sense, etc.). The evaluation of high-stakes writing projects can be switched such that students compose a metacognitive reflection of their mindfulness experience while developing a project (of any genre or about any subject), and it's that reflection that is graded. I want to be clear here that students who struggle to understand mindfulness are not penalized if they don't become as "enlightened" as the student at the next computer—although every student I've worked with has eventually caught more than a glimpse of mindfulness.

PROPOSED AMENDMENTS TO WRITING POLICY OUTCOMES

- Examination of the interplay of formlessness and form throughout the writing process and in multiple rhetorical situations.
- Increased instructional time allotted to writing experiences that yield fragmentary and partial compositions as a way to disrupt assumptions about the writing process, student writers, and readers.
- Extended experiences in prewriting that may or may not yield more polished compositions with discussion of embodied and affective writing experiences such as felt sense; an equal or higher ratio of incomplete to complete texts.
- Developing appreciation of nonwriting as a necessary component of writing experience.

Alleviating the Suffering of Writers

Those of us interested in reducing the cycle of suffering from writing instruction are on the path to becoming bodhisattva of writing. Bodhisattva respond to the call to realign themselves to the present moment and to renounce types of instruction that obstruct, ignore, or diminish present awareness. In traditional Indian Buddhism, practitioners devoted to the spiritual path took one of three directions: they might become *śrāvaka* or disciples who attended the teachings of the Buddha and would possibly exit the cycle of suffering by becoming an *arhat* at death; or they might become a *pratyekabuddha*, a monk who practiced in solitude and was not directly exposed to the Buddha's teachings during the lifetime that ends in their enlightenment; or they might become *bodhisattva*, an individual who achieves enlightenment but delays his or her release into nirvana in order to help others (Lopez 65). Such a practitioner has a compassionate mind of *bodhicitta*, which strives for personal awakening in order to help other sentient beings achieve enlightenment and freedom from the cycle of suffering. As Donald Lopez describes a *bodhisattva*, this is the "extraordinary person" who will discover "the path to nirvāṇa through his own efforts and teach it to the world" (67). The chief example of a *bodhisattva* was the Buddha, who stuck around for decades after his enlightenment under the bodhi tree. As Pema Chödrön describes the motivation of the bodhisattva, "Few of us are satisfied with retreating from the world and just working on ourselves. We want our training to manifest and to be of benefit. The bodhisattva-warrior, therefore, makes a vow to wake up not just for himself but for the welfare of all beings" (*The Places* 93). In many regards, the field of composition studies in my country of residence, the United States, is already headed in this direction.

We are a group of professionals who have dedicated years of study and effort to helping others improve the paramount human activity of expressing self and communicating with society, and our adjunct colleagues most embody this compassion. The sheer depth and range of theoretical approaches and pedagogical applications offering insight into writing matters as differentiated as genre, code meshing, paragraph structures, freewriting, connotative language, collaboration, embodied rhetoric, narrative, and the Toulmin method are commendable. And yet dislike, antipathy, self-recrimination, and avoidance dominate the views on writing of innumerable individuals, despite our best efforts, negative views which persist long after graduation. The impression is that writing suffering is impossible to change, a "necessary evil" of the work of a student and the profession of teaching, and maybe because we ourselves didn't have a particularly easy path, we were trained under teachers of mindlessness, we lack role models. Maybe it's because we have been hurt as writers by the instruction of mindlessness.

Rather than sweeping it under the carpet, as bodhisattva of mindful writing, we take a long critical look at writing suffering and how our own students' suffering might be a direct result of our future-oriented rhetorics. A bodhisattva of mindful writing endeavors to help first-year composition students dwell in the present moment for the purposes of writing by developing a present-based rhetorical awareness of metacognition and non-evaluation that is receptive of intrapersonal rhetoric and impermanence. Every writing-related present moment is of deep interest and defines us as teachers. Mary Rose O'Reilley proposes that "we shift our sense of ourselves as successful teachers away from the quality of our corrections and toward the quality of our mindfulness" (*Peaceable* 49). A bodhisattva of composition makes open inquiry into his or her teaching approach, examining the entrenched "pedagogical narrative" or preconceptions about what it means to learn, write, or teach (Brown 80–81). As instructors, we need to remind ourselves, as Sid Brown does in *The Buddhist in the Classroom*, "that students and teachers can unwittingly cause each other great harm, perhaps especially when their actions are not motivated by generosity, loving-kindness, compassion, and altruistic joy, informed by clarity and understanding" (80–81). Moreover, if you yourself suffer as a writer and do not address it, you will invariably pass suffering onto your students. Mindlessness in writing undermines confidence, sets in stone lifelong attitudes about writing and judgments about self-efficacy, and ossifies these views into career choices. If we do not address the harms of mindlessness in our writing, these negative attributes are transmitted to other people—to our students, colleagues, and children. It's like the pre-flight instruction admonishing adult passengers to take care of their own oxygen masks before assisting others: composition teachers' mindful instruction necessitates their own mindful writing. Mindlessness in writing perpetuates mindlessness in other people's lives.

Thus as bodhisattva of writing, we 1) acknowledge the suffering caused by mainstream writing education through its future-oriented rhetorics and praxis; 2) endeavor to realign each occasion of writing instruction and writing experience so that it more fully occupies a present moment; and 3) accept that customary instructional experiences will be replaced or joined by ones with less certainty and less control as we move beyond a pursuit of fixed outcomes, polished works, and kowtowing to hypothetical audiences who dwell in an imagined future. We swap out the traditional certainties that come with future-bound rhetorics (and come with high costs of mindlessness) for the bounties of impermanence and interconnection. We seek out the abundancy of intrapersonal rhetoric, knowing that it comes with messes and imperfections; we welcome the non-control of the nonverbal, preverbal, and prewriting, the fragmentary and the elliptical, that which trails off and is left unfinished. Conversations will be unusual, at times uncomfortable, strange, and unprecedented for both the teacher and the student. We make these changes because achieving writing calm, equanimity, and a non-evaluative fluency are far more valuable learning outcomes than a polished rhetorical analysis.

To change how students think of writing is no small task, but it can be a deeply moving, deeply eye-opening experience for the teacher to see students work in this new approach. Endless energy is expended on reaching after standards, goals, always compelled to be a different kind of writer, to aspire, to abandon their current situation for chimerical future prospects. To let go of all of that effort and *simply, finally*, be content with the writing and teaching and learning moment that we have in our midst is invigorating. We don't need to chase after anything about writing or summon energy to inspire our students to make this aspirational chase. We can stop asking students to abandon their current writing selves for seemingly better writing selves. As teachers, what a relief not to be yet another representative of that restlessness, that dissatisfaction, that discontent and doubt.

How can we become mindful writing instructors inside a mindless institution? In one way, it only takes the moment-by-moment, lived mindfulness of the instructor to change things. By remaining attuned to the present moment, during the hours we teach, during curriculum and course design, during office hours, during department meetings, and in our own writing practice, we can reduce writing suffering. It's both as simple and as demanding as that: eyes on the moment. It can start with us in the next moment, the very next moment of teaching. Right now. We don't require special circumstances, and as this book has hopefully proved, mindfulness dovetails neatly into rhetorical situation, process as a chronology, regular elements of writing instruction. This very moment is perfect; (to return to Pema Chödrön) "[t]his very moment is the perfect teacher"; and in this very moment, we are perfect as we are as writing teachers (*When* 12). Even a single impactful assignment or class discussion that enables a student to settle into the present and write from within a moment can lead to long-term differences. Mindful composition pedagogy demonstrates how this happy occasion is not merely a fluke but caused by choices made by student and instructor, an outlook that student writers can maintain in other courses and throughout their writing lives. As the Buddha said in the Diamond Sutra, if bodhisattvas "kept in mind any such arbitrary conceptions as one's own self, other selves, living beings, or a universal self, they would not be called Bodhisattva ... It means that there are no sentient beings to be delivered and there is no selfhood that can begin the practice of seeking to attain [enlightenment]" (Goddard 97–98). The Buddha applied this own non-specialness to himself; asked by the disciple Subhuti whether he felt he had acquired something special upon the moment of his own enlightenment, the Buddha explained that what he had experienced "is the same as what all others have attained," something "undifferentiated, neither to be regarded as a high state, nor is it to be regarded as a low state. It is wholly independent of any definitive or arbitrary conceptions of any individual self" (Goddard 104–105). This mindfulness transformation is as basic as you, your students, and this moment in the recognition that the ability to write is always present.

May each moment of your breathing
Be a field of asterisk-words, word-flowers.
May your writing
Be free of suffering.
May your writing
Be the cessation of suffering.
May your writing instruction
Be free of suffering.
May your writing instruction
Be the cessation of suffering.

BRIEF GLOSSARY

Anatta: no-self or without ego; individuals are empty because devoid of independent existence

Anicca: impermanence; existence is transient and ever-changing

Bodhicitta: a state of mind which seeks its own enlightenment and the enlightenment of others

Bodhisattva: a person who has devoted his or her life to achieving enlightenment and to relieving the suffering of other people as he or she strives toward enlightenment

Citta: consciousness of mind and feelings; mood or outlook as it changes from moment to moment

Dharma: the teachings of the Buddha in doctrine and practice

Dukkha: suffering from unsatisfactory conditions of life

Eightfold Path: the fourth part of the Buddha's Four Noble Truths; an explication of eight steps to relinquish suffering and achieve enlightenment: Right Understanding, Right Mindedness, Right Speech, Right Action, Right Living, Right Effort, Right Attentiveness, and Right Concentration

Four Noble Truths: the process explained by the Buddha for a release from suffering

Gautama: the birth name of the individual of royal birth who became the Buddha sometime between the fourth and the sixth century BCE in India

Groundlessness: the condition of constant impermanence without discrete and lasting entities or phenomena

Interbeing: the view that the self contains non-self elements

Intrapersonal rhetoric: consistently occurring discursive mental formations; self-talk; voice

Maitri: calm, non-reactive mindset to external developments

Monkey mind: intrapersonal rhetoric that operates wildly in an individual's mind without monitoring, causing the individual to neglect the present moment; frequently results in elaborate preconceptions and storylines

Prajñāpāramitā: the perfection of wisdom that can arise from the contemplation of emptiness

Preconception: a type of storyline that specifically concerns the future; an assumption about events occurring in a temporal setting that is not the moment at hand

Right Attentiveness: also called Right Mindfulness; the seventh step on the Eightfold Path to enlightenment; entails contemplation of the body, feelings, mind, and phenomena

Right Discipline: also called Right Effort; the sixth step on the Eightfold Path to enlightenment; entails steps to avoid and overcome negative mental formations that cause suffering and the development of positive mental formations characteristic of enlightenment, such as attentiveness, tranquility, and concentration

Samādhi: a still mind and state of deep concentration; meditative consciousness

Saṃsāra: a form of suffering from remaining in the cycles of birth and death that results from attachment and false views

Sangha: a community of Buddhist or mindfulness practitioners

Skandha: the five physical and mental components of a person: form, feeling, discrimination, conditioning factors, and consciousness

Storyline: an elaborate fantasy narrated by intrapersonal rhetoric in which a practitioner abandons present awareness for past- or future-set illusions

Sūnyatā: the emptiness in which all entities and phenomena, including the human ego, lack independent existence; the repudiation of a particular kind of existence (independent and permanent)

Sūtra: a teaching of the Buddha or of one of his sanctioned disciples

BIBLIOGRAPHY

Anālayo. *Perspectives on Satipaṭṭhāna*. Cambridge: Windhorse Publications, 2013.

Aristotle. "From Rhetoric." *The Rhetorical Tradition: Readings from Classical Times to the Present*, edited by Patricia Bizzell and Bruce Herzberg, 3rd ed., Boston: Bedford / St. Martin's, 2001.

Ashbery, John. *Selected Poems*. New York: Penguin, 1985.

Atwill, Janet M., and Janice M. Lauer, editors. *Perspectives on Rhetorical Invention*. Knoxville, TN: University of Tennessee Press, 2002.

Baer, Ruth A. "Measuring Mindfulness." *Contemporary Buddhist*, vol. 12, no. 1, 2011, pp. 241–261.

Bakhtin, M. M. *The Dialogic Imagination*. Austin, TX: University of Texas Press, 1981.

Barbezat, Daniel P. and Mirabai Bush. *Contemplative Practices in Higher Education: Powerful Methods to Transform Teaching and Learning*. San Francisco, CA: Jossey-Bass, 2014.

Bartholomae, David. "Writing with Teachers: A Conversation with Peter Elbow." *College Composition and Communication*, vol. 46, no. 1, 1995, pp. 62–71.

Batchelor, Stephen. *Secular Buddhism: Imagining the Dharma in an Uncertain World*. New Haven, CT: Yale University Press, 2017.

Berila, Beth. *Integrating Mindfulness into Anti-Oppression Pedagogy: Social Justice in Higher Education*. New York: Routledge, 2016.

Berlin, James A. "Rhetoric and Ideology in the Writing Class." *College English*, vol. 50, no. 5, 1988, pp.477–494.

—. *Rhetoric and Reality: Writing Instruction in American Colleges, 1900–1985*. Carbondale, IL: Southern Illinois University Press, 1987.

Biesecker, Barbara A. "Rethinking the Rhetorical Situation from within the Thematic of "Différance."" *Philosophy & Rhetoric*, vol. 22, no. 2, 1989, pp. 110–130.

Bitzer, Lloyd F. "The Rhetorical Situation." *Philosophy & Rhetoric*, vol. 1, 1968, pp. 1–14.

Bizzell, Patricia. "Cognition, Convention, and Certainty: What We Need to Know about Writing." *PRE/TEXT*, vol. 3, no/ 3, 1982, pp. 213–243.

Bohm, David. *On Dialogue*. London: Routledge, 1996.

Boice, Robert "Writerly Rules for Teachers." *The Journal of Higher Education*, vol. 66, no. 1, 1995, pp. 32–60.

—. *How Writers Journey to Comfort and Fluency: A Psychological Adventure*. Westport, CT: Praeger, 1994.
—. "Writing Blocks and Tacit Knowledge." *The Journal of Higher Education*, vol. 64, No. 1, 1993, pp. 19–54.
—. "Combining Writing Block Treatments: Theory and Research." *Behavior Research and Therapy*, vol. 30, no. 2, 1992, pp. 107–116.
—. "Cognitive Components of Blocking." *Written Communication*, vol. 2, no. 1, 1985, pp. 91–104.
—. "The Neglected Third Factor in Writing: Productivity." *College Composition and Communication*, vol. 36, no. 4, 1985, pp. 472–480.
Boice, Robert and Ferdinand Jones. "Why Academicians Don't Write." *The Journal of Higher Education*, vol. 55, no. 5, 1984, pp. 567–582.
Boice, Robert and Patricia E. Meyers. "Two Parallel Traditions: Automatic Writing and Free Writing." *Written Communication*, vol. 3, no. 4, 1986, pp. 471–490.
Booth, Wayne C. "The Rhetorical Stance." *College Composition and Communication*, vol. 14, no, 3, 1963, pp.139–145.
Brand, Alice. "Social Cognition, Emotions, and the Psychology of Writing." *Journal of Advanced Composition*, vol. 11, no. 2, 1991, pp. 395–407.
—. "The Why of Cognition: Emotion and the Writing Process." *College Composition and Communication*, vol. 38, no. 4, 1987, pp. 436–443.
Brand, Alice Glarden and Richard L. Graves. *Presence of Mind: Writing and the Domain beyond the Cognitive*. Portsmouth, NH: Boynton/Cook Heinemann, 1994.
Breuch, Lee-Ann M. Kastman. "Post-Process 'Pedagogy': A Philosophical Exercise." *Journal of Advanced Composition*, vol. 22, no. 1, 2002, pp. 119–150.
Britton, James. "Spectator Role and the Beginnings of Writing." *Cross – Talk in Comp Theory: A Reader*, edited by Victor Villanueva, Urbana: NCTE, 2003, pp. 151–174.
—. "Shaping at the Point of Utterance." *Landmark Essays on Rhetorical Invention and Writing*, edited by Richard E. Young and Yameng Liu, New York: Routledge, 1994, pp. 147–152.
Brodkey, Linda. "Modernism and the Scene(s) of Writing." *College English*, vol. 49, no. 4, 1987, pp. 396–418.
Brown, Sid. *A Buddhist in the Classroom*. New York: State University of New York Press, 2008.
Brummett, Barry. "Rhetorical Epistemology and Rhetorical Spirituality." *The Academy and the Possibility of Belief: Essays on Intellectual and Spiritual Life*, edited by Mary Louise Buley-Meissner, Mary McCaslin Thompson, and Elizabeth Bachrach Tan, New York: Hampton Press, 2000, pp. 121–135.
Bruning, Roger and Christy Horn. "Developing Motivation to Write." *Educational Psychologie*, vol. 35, no. 12,000, pp. 25–37.
Bush, Mirabai. "Mindfulness in Higher Education." *Contemporary Buddhism: An Interdisciplinary Journal*, vol. 12, no. 11, 2011, pp. 183–197.
Campbell, JoAnn. "Writing to Heal: Using Meditation in the Writing Process." *College Composition and Communication*, vol. 45, 1994, pp. 246–251.
Carillo, Ellen C. *Securing a Place for Reading in Composition: The Importance of Teaching for Transfer*. Utah: Utah State University Press, 2015.
Carroll, Lee Ann. *Rehearsing New Roles: How College Students Develop as Writers*. Carbondale, IL: Southern Illinois University Press, 2002.
Carson, Shelley H. and Ellen J. Langer. "Mindfulness and Self-Acceptance." *Journal of Rational-Emotive & Cognitive-Behaviour Therapy*, vol. 24, no. 1, 2006, pp. 29–43.
Cassidy, Laurie. "Mindful Breathing: Creating Counterpublic Space in the Religious Studies Classroom." *Journal of Feminist Studies in Religion*, vol. 28, no. 1, pp. 164–177.

Chanowitz, Benzion, and Ellen J. Langer. "Premature Cognitive Commitment." *Journal of Personality and Social Psychology*, vol. 41, no. 6, 1981, pp. 1051–1063.

Chödrön, Pema. *How to Meditate: A Practical Guide to Making Friends with Your Mind*. Louisville, CO: Sounds True, 2013.

—. *The Places That Scare You: A Guide to Fearlessness in Difficult Times*. Boulder, CO: Shambhala, 2001.

—. *When Things Fall Apart: Heart Advice for Difficult Times*. Boulder, CO: Shambhala, 2000.

—. *The Wisdom of No Escape and the Path of Loving-Kindness*. Boulder, CO: Shambhala, 1991.

Cohen, Judith Bell. "The Missing Body—Yoga and Higher Education." *Journal of the Assembly for Advanced Perspectives on Learning*, vol. 12, 2006–2007, pp. 4–24.

Collins, Steven. *Selfless Persons: Imagery and Thought in* Theravāda *Buddhism*. New York: Cambridge University Press, 1982.

Commons, Michael Lamport, and Dristi Adhikari. "Possible Components of Mindfulness." *Critical Mindfulness: Exploring Langerian Models*, edited by Sayyed Mohsen Fatemi, New York: Springer, 2016, pp. 193–202.

Conference on College Composition and Communication. "Principles for the Postsecondary Teaching of Writing," 2015, www.ncte.org/cccc/resources/positions/postsecondarywriting.

Consilio, Jennifer. "Making MAC (Mindfulness Across the Curriculum) Happen: Creating a Mindful Campus Culture." *College Conference of Composition and Communication*, Portland, Oregon, 2017.

Consigny, Scott. "Rhetoric and Its Situations." *Philosophy & Rhetoric*, vol. 7, no. 3, 1974, pp. 175–186.

Conze, Edward. *Buddhist Wisdom: The Diamond Sutra and The Heart Sutra*. London: Vintage, 2001.

Council of Writing Program Administrators. "WPA Outcomes Statement for First-Year composition V 3.0," 2014, http://wpacouncil.org/positions/outcomes.html/.

Council of Writing Program Administrators, National Council of Teachers of English and National Writing Project. "Framework for Success in Postsecondary Writing," 2011.

Crawford, Ilene. "Building a Theory of Affect in Cultural Studies Composition Pedagogy." *Journal of Advanced Composition*, vol. 22, no. 3, 2002, pp. 678–684.

Crawford, Ryan, and Andreas Wilhoff. "Stillness in the Composition Classroom: Insight, Incubation, Improvisation, Flow, and Meditation." *JAEPL*, vol. 19, 2013–2014, pp. 74–83.

Crick, Nathan. "Composition as Experience: John Dewey on Creative Expression and the Origins of 'Mind.'" *College Composition and Communication*, vol. 55, no. 2, 2003, pp. 254–275.

Csikszentmihalyi, Mihaly. *Flow: The Psychology of Optimal Experience*. London: Harper Perennial, 1990.

Daly, John A. "Writing Apprehension." *When a Writer Can't Write*, edited by Mike Rose, New York: Guilford Press, 1985, pp. 43–82.

Daly, John A., and Wayne G. Shamo. "Writing Apprehension and Occupational Choice." *Journal of Occupational Psychology*, vol. 49, 1976, pp. 55–56.

D'Angelo, Frank. "The Rhetoric of Intertextuality." *Rhetoric Review*, vol. 29, no. 1, 2010, pp. 31–47.

DeLuca, Geraldine. "Headstands, Writing, and the Rhetoric of Radical Self-Acceptance." *Journal of the Assembly for Advanced Perspectives on Learning*, vol. 11, 2005–2006, pp. 27–41.

De Silva, Padmasiri. "Theoretical Perspectives on Emotions in Early Buddhism." *Emotions in Asian Thought: A Dialogue in Comparative Philosophy*, edited by J. Marks and Roger Ames, New York: State University Press of New York, 1995, pp. 109–121.

Dillon, George L. *Rhetoric as Social Imagination: Explorations in the Interpersonal Function of Language*. Indiana: Indiana University Press, 1986.

Dirk, Kerry. "Navigating Genres." *Writing Spaces: Readings on Writing, Volume 1*, Parlor Press, 2010, http://writingspaces.org/sites/default/files/dirk--navigating-genres.pdf.

Ede, Lisa, and Andrea Lunsford. "Representing Audience: 'Successful' Discourse and Disciplinary Critique." *Engaging Audience: Writing in an Age of New Literacies*, edited by M. Elizabeth Weiser, Brian M. Fehler, and Angela M. González, Urbana: NCTE, 2009, pp. 26–41.

—. "Audience Addressed/Audience Invoked: The Role of Audience." *Cross-Talk in Comp Theory: A Reader*, edited by Victor Villanueva, Urbana: NCTE, 2003, pp. 77–95.

Elbow, Peter. *Vernacular Eloquence: What Speech Can Bring to Writing*. Oxford: Oxford University Press, 2012.

—. "Closing My Eyes as I Speak." *Everyone Can Write: Essays toward a Hopeful Theory of Writing and Teaching Writing*. Oxford: Oxford University Press, 2000, pp. 93–112.

—. "In Defense of Private Writing: Consequences for Theory and Research." *Everyone Can Write: Essays toward a Hopeful Theory of Writing and Teaching Writing*. Oxford: Oxford University Press, 2000, pp. 257–280.

—. "Toward a Phenomenology of Freewriting." *Everyone Can Write: Essays toward a Hopeful Theory of Writing and Teaching Writing*. Oxford: Oxford University Press, 2000, pp. 113–136.

—. *Writing with Power*. Oxford: Oxford University Press, 1998.

—. "Silence: A Collage." *Presence of Mind: Writing and the Domain beyond the Cognitive*, edited by Alice Glarden Brand and Richard L. Graves, Portsmouth, NH: Boynton/Cook Heinemann, 1994, pp. 9–20.

—. *Embracing Contraries: Explorations in Learning and Teaching*. Oxford: Oxford University Press, 1986.

Edbauer, Jenny. "Unframing Models of Public Distribution: From Rhetorical Situation to Rhetorical Ecologies." *Rhetoric Society Quarterly*, vol. 25, no. 4, 2005, pp. 5–24.

Ekman, Paul, et al. "Buddhist and Psychological Perspectives on Emotions and Well-Being." *Current Directions in Psychological Science*, vol. 14, no. 2, 2005, pp. 59–72.

Emig, Janet. "The Uses of the Unconscious in Composing." *College Composition and Communication*, vol. 15, 1964, pp. 6–11.

Enos, Richard L. "Inventional Constraints on Technographers of Ancient Athens: A Study of Kairos." *Rhetoric and Kairos: Essays in History, Theory, and Praxis*, edited by Phillip Sipiora and James S. Baumlin, New York: State University of New York Press, 2002, pp. 77–88.

Eubanks, Philip. "Poetics and Narrativity: How Texts Tell Stories." *What Writing Does and How It Does It: An Introduction to Analyzing Texts and Textual Practices*, edited by Charles Bazerman and Paul Prior, New York: Lawrence Erlbaum, 2004.

Evans, David. "Gratitude." *Buddhist Poetry Review*, vol. 2, no. 2.

Feldenkrais, Moshé. *Awareness through Movement: Health Exercises for Personal Growth*. New York: Harper & Row, 1977.

Feleppa, Robert. "Zen, Emotion, and Social Engagement." *Philosophy East & West*, vol. 59, no. 3, 2009, pp. 263–293.

Fink, Charles K. "The 'Scent' of a Self. Buddhism and the First-Person Perspective." *Asian Philosophy*, vol. 22, no. 3, 2012, pp. 289–306.

Flaherty, Alice W. *The Midnight Disease: The Drive to Write, Writer's Block, and the Creative Brain*. Boston: Houghton Mifflin, 2004.

Flavell, John H. "Metacognition and Cognitive Monitoring: A New Area of Cognitive-Developmental Inquiry." *American Psychologist*, vol. 34, no. 10, 1992, pp. 906–911.

Fleckenstein, Kristie S. *Embodied Literacies: Imageword and a Poetics of Teaching*. Carbondale, IL: Southern Illinois University Press, 2003.

—. "Writing Bodies: Somatic Mind in Composition Studies." *College English*, vol. 61, no. 3, 1999, pp. 281–206.

Fleury, Staci and Alexandria Peary. "The Montaigne Method: Adding Content and Consciousness Through Revision as Invention." *Journal of Teaching Writing*, Vol. 29, No. 2, 2014, pp. 1–14.

Flower, Linda. "Cognition, Context, and Theory Building." *College Composition and Communication*, vol. 40, no. 3, 1989, pp. 282–311.

—. "Writer-Based Prose: A Cognitive Basis for Problems in Writing." *College English*, vol. 41, 1979, pp. 19–37.

Flower, Linda, and John R. Hayes. "A Cognitive Process Theory of Writing." *College Composition and Communication*, vol. 32, no. 4, 1981, pp. 365–387.

Fontaine, Sheryl I. "Teaching with the Beginner's Mind: Notes from My Karate Journal." *College Composition and Communication*, vol. 54, no. 2, 2002, pp. 208–221.

Forrester, Rachel. "The 'Not Trying' of Writing." *JAEPL*, vol. 13, 2007–2008, pp. 45–56.

Gallehr, Donald R. "Wait, and the Writing Will Come: Meditation and the Composing Process." *Presence of Mind: Writing and the Domain beyond the Cognitive*, edited by Alice Glarden Brand and Richard L. Graves, Portsmouth, NH: Boynton/Cook Heinemann, 1994, pp. 21–30.

Gardner, Howard. *Frames of Mind: The Theory of Multiple Intelligences*. New York: Basic Books, 2011.

Garfield, Jay L. *The Fundamental Wisdom of the Middle Way: Nāgārjuna's* Mūlamadhyamakakārikā. Oxford: Oxford University Press, 1995.

Gendlin, Eugene. *Experiencing and the Creation of Meaning: A Philosophical and Psychological Approach to the Subjective*. Evanston, IL: Northwestern University Press, 1962.

Gethin, Rupert. "On Some Definitions of Mindfulness." *Contemporary Buddhism*, vol. 12, no. 1, 2011, pp. 263–279.

—. *The Foundations of Buddhism*. New York: Oxford University Press, 1998.

Ghiselin, Brewster. *The Creative Process*. New York: New American Library, 1952.

Glenn, Cheryl. *Unspoken: A Rhetoric of Silence*. Carbondale, IL: Southern Illinois University Press, 2004.

Glenn, Cheryl and Krista Ratcliffe, editors. *Silence and Listening as Rhetorical Acts*. Carbondale, IL: Southern Illinois University Press, 2011.

Goddard, Dwight, editor. *A Buddhist Bible*. Boston: Beacon Press, 1994.

Gombrich, Richard. *What the Buddha Thought*. Sheffield: Equinox, 2009.

Grant, Adam M., et al. "Mindful Creativity: Drawing to Distinctions." *Creativity Research Journal*, vol. 16, nos. 2 and 3, 2004, pp. 261–265.

Grant-Davie, Keith. "Rhetorical Situations and Their Constituents." *Rhetoric Review* vol. 15, no. 2, 1997, pp. 264–279.

Grossman, Paul, and Nicholas T. Van Dam. "Mindfulness by Any Other Name …: Trials and Tribulations of *Sati* in Western Psychology and Science." *Contemporary Buddhism*, vol. 12, no. 1, pp. 219–239.

Gunaratana, Bhante Henepola. *Mindfulness in Plain English*. Somerville, MA: Wisdom Publications, 2002.

Gunaratne, Shelton A., Mark Pearson, and Sugath Senarath. *Mindful Journalism and News Ethics in the Digital Era: A Buddhist Approach*. New York: Routledge, 2015.

Hanh, Thich Nhat. *Silence: The Power of Quiet in a World Full of Noise*. New York: Harper Collins, 2015.

—. *Anger: Wisdom for Cooling the Flames*. New York: Penguin Random House, 2001.

—. *The Path of Emancipation*. Berkeley, CA: Parallax Press, 2000.

—. *The Heart of the Buddha's Teaching*. New York: Broadway Books, 1999.

Harding, D.W. *Experience into Words*. New York: Horizon Press, 1964.

Hart, Tobin. "Opening the Contemplative Mind in the Classroom." *Journal of Transformative Education*, vol. 2, no. 1, 2004, pp. 28–46.

Hershock, Peter D. "Renegade Emotion: Buddhist Precedents for Returning Rationality to the Heart." *Philosophy East & West*, vol. 53, no. 2, 2003, pp. 251–270.

Hesse, Doug. "Number of FYC courses taught each year?" *WPA list-serv*. WPA-L@asu.edu. Accessed 23 Jan. 2012.

Hjortshoj, Keith. *Understanding Writing Blocks*. New York: Oxford University Press, 2001.

Hyland, Terry. *Mindfulness and Learning: Celebrating the Affective Dimension of Education*. New York: Springer, 2011.

Inoue, Asao B. *Antiracist Writing Assessment Ecologies: Teaching and Assessing Writing for a Socially Just Future*. Anderson, SC: Parlor Press, 2015.

—. "A Grade-Less Writing Course That Focuses on Labor and Assessing." *First-Year Composition: From Theory to Practice*, edited by Deborah Coxwell Teague and Ronald F. Lunsford, Anderson, SC: Parlor Press, 2014, pp. 71–110.

James, William. *The Principles of Psychology*. Vol. 1. 1890. New York: Dover Publications, 1918.

Johnson, Kristine. "Beyond Standards: Disciplinary and National Perspectives on Habits of Mind." *College Composition and Communication*, vol. 64, no. 3, 2013, pp. 517–541.

Johnstone, Keith. *Improv: Improvisation and the Theater*. New York: Routledge, 1989.

Jones, Donald C. "Beyond the Postmodern Impasse of Agency: The Resounding Relevance of John Dewey's Tacit Tradition." *JAC*, vol. 16, no. 1, 1996, pp. 81–102.

Kabat-Zinn, Jon. *Wherever You Go, There You Are: Mindfulness Meditation in Everyday Life*. London: Hachette Books, 1994.

Kalamaras, George. "East Meets West: Peter Elbow's 'Embracing' of 'Contraries' Across Cultures." *Writing with Elbow*, edited by Pat Belanoff et al., Utah: Utah State University Press, 2002, pp. 114–124.

—. "Meditative Silence and Reciprocity: The Dialogic Implications for 'Spiritual Sites of Composing.'" *JAEPL*, vol. 2, 1996–1997, pp. 18–26.

—. *Reclaiming the Tacit Dimension: Symbolic Form in the Rhetoric of Silence*. New York: State University of New York Press, 1994.

"Kassaka Sutra: The Farmer." Translated from the Pali by Thanissaro Bhikkhu. *Access to Insight (Legacy Edition)*, www.accesstoinsight.org/tipitaka/sn/sn04/sn04.019.than.html.

Kasulis, Thomas P. *Zen Action, Zen Person*. Honolulu, HI: University of Hawaii Press, 1981.

Kearney, Julie. "Writing as an Altered State of Consciousness: Process, Pedagogy, and Spirituality." *JAEPL*, vol. 16, 2010–2011, pp. 67–78.

Kellogg, Ronald T. *The Psychology of Writing*. New York: Oxford University Press, 1994.

Kent, Thomas. *Paralogic Rhetoric: A Theory of Communicative Interaction*. Lewisburg, PA: Bucknell University Press, 1993.

Kern, Stephen. *The Culture of Time and Space:1880–1918*. Cambridge, MA: Harvard University Press, 1983.

Kerr, Tom. "The Feeling of What Happens in Departments of English." *A Way to Move: Rhetorics of Emotion & Composition Studies*, edited by Dale Jacobs and Laura R. Micciche, Portsmouth, NH: Boynton/Cook Heinemann, 2003, pp. 23–32.

Khema, Ayya. *Who Is My Self?: A Guide to Buddhist Meditation*. Somerville, MA: Wisdom Publications, 1997.

Kinneavy, James L. "Kairos: A Neglected Concept in Classical Rhetoric." *Rhetoric and Praxis: The Contribution of Classical Rhetoric to Practical Reasoning*, edited by Jean Dietz Moss, Washington D.C.: Catholic University Press of America, 1986, pp. 79–105.

Kinneavy, James L. and Catherine R. Eskin. "Kairos in Aristotle's Rhetoric." *Written Communication*, vol. 17. no. 3, 2000, pp. 432–444.

Kirsch, Gesa. "Creative Spaces for Listening, Learning, and Sustaining the Inner Lives of Students." *JAEPL*, vol. 14, 2009, pp. 56–67.

—. "From Introspection to Action: Connecting Spirituality and Civic Engagement." *College Composition and Communication*, vol. 60, no. 4, 2009, W1–W15.

Kirsch, Gesa E. and Joy S. Ritchie. "Beyond the Personal: Theorizing a Politics of Location in Composition Research." *College Composition and Communication*, vol. 46, no. 1, 1995, pp. 7–29.

Knoblauch, A. Abby. "Bodies of Knowledge: Definitions, Delineations, and Implications of Embodied Writing in the Academy." *Composition Studies*, vol. 40, no. 2, 2012, pp. 50–65.

Kroll, Barry M. *The Open Hand: Arguing as an Art of Peace*. Utah: Utah State University Press, 2013.

Lakoff, George, and Mark Johnson. *Metaphors We Live By*. Chigaco: University of Chicago Press, 1980.

Langer, Ellen J. *The Power of Mindful Learning*. Boston, MA: Da Capo Press, 1997.

—. "A Mindful Education." *Educational Psychologist*, vol. 28, no. 1, 1993, pp. 43–50.

—. *Mindfulness*. Boston, MA: Da Capo Press, 1989.

Langer, Ellen J., and Alison I. Piper. "The Prevention of Mindlessness." *Journal of Personality and Social Psychology*, vol. 53, no, 2, pp. 280–287.

Langer, Ellen J., and Mihnea Moldoveanu. "The Construct of Mindfulness." *Journal of Social Issues*, vol. 56, no. 1, 2000, pp. 1–9.

Langstraat, Lisa. "The Point Is There Is No Point: Miasmic Cynicism and Cultural Studies Composition." *Journal of Advanced Composition*, vol. 2, no. 2, 2002, pp. 293–325.

Larson, Reed. "Emotional Scenarios in the Writing Process: An Examination of Young Writers' Affective Experiences." *When a Writer Can't Write*, edited by Mike Rose, New York: Guilford Press, 1985, pp. 19–42.

Lauer, Janice M. *Invention in Rhetoric and Composition*. Anderson, SC: WAC Clearinghouse/ Parlor Press, 2004.

LeFevre, Karen Burke. *Invention as a Social Act*. Carbondale, IL: Southern Illinois University Press, 1987.

LeMesurier, Jennifer Lin. "Somatic Metaphors: Embodied Recognition of Rhetorical Opportunities." *Rhetoric Review*, vol. 33, no. 4, 2014, pp. 362–380.

Lofty, John S. *Time to Write: The Influence of Time and Culture on Learning to Write*. 2nd ed., New York: State University of New York, 2015.

Lopez, Donald S. *The Story of Buddhism: A Concise guide to Its History & Teachings*. San Francisco: Harper Collins, 2001.

Lovejoy, Kim Brian. "Self-Directed Writing: Giving Voice to Student Writers." *English Journal*, vol. 98, no, 6, 2009, pp. 79–86.

Low, Sor-Ching. "Romancing Emptiness." *Contemporary Buddhism*, vol. 7, no. 2, 2006, pp. 129–147.
Loy, David R. "Awareness Bound and Unbound: Realizing the Nature of Attention." *Philosophy East and West*, vol. 58, no. 2, 2008, pp. 223–243.
Lyons, Tim. "Bridging the Gap between Nāgārjuna and the University Rhetoric Class: A Guide with Links to Assignments." *ETC*, 2010, pp. 328–347.
Maitra, Keya. "The Questions of Identity and Agency in Feminism without Borders: A Mindful Response." *Hypatia*, vol. 28, no. 2, 2013, pp. 360–376.
Mandel, Barrett J. "The Writer Writing Is Not at Home." *College Composition and Communication*, vol. 31, no. 4, 1980, pp. 370–377.
—. "Losing One's Mind: Learning to Write and Edit." *College Composition and Communication*, vol. 29, no. 4, 1978, pp. 362–368.
Mathieu, Paula. "Being There: Mindfulness as Ethical Classroom Practice." *JAEPL*, vol. 21, 2015–2016, pp. 14–20.
McDonald, Kathleen. *How to Meditate: A Practical Guide*. Somerville, MA: Wisdom Publications, 2005.
McLeod, Susan H. "The Affective Domain and the Writing Process: Working Definitions." *Journal of Advanced Composition*, vol. 11, no. 1, 1991, pp. 95–105.
—. "Some Thoughts about Feelings: The Affective Domain and the Writing Process." *College Composition and Communication*, vol. 38, no. 4, 1987, pp. 426–435.
Meyer, Ulrich. *The Nature of Time*. Oxford University Press, 2013.
Micciche, Laura R. *Doing Emotion: Rhetoric, Writing, Teaching*. Portsmouth, NH: Boynton/Cook Heinemann, 2007.
Miller, Carolyn R. "Genre as Social Action." *Quarterly Journal of Speech*, vol. 70, no. 2, 1984, pp. 151–167.
Moffett, James. "Liberating Inner Speech." *College Composition and Communication*, vol. 36, no. 3, 1985, pp. 304–308.
—. "Reading and Writing as Meditation." *Language Arts*, vol. 60, no. 3, 1983, pp. 315–322, 332.
—. "Writing, Inner Speech, and Meditation." *College English*, vol. 44, no. 3, 1982, pp. 231–246.
—. "Integrity in the Teaching of Writing." *The Phi Delta Kappa*, vol. 61, no. 4, 1979, pp. 276–279.
Moon, Gretchen Flesher. "The Pathos of *Pathos*: The Treatment of Emotion in Contemporary Composition Textbooks." A *Way to Move: Rhetorics of Emotion & Composition Studies*, edited by Dale Jacobs and Laura R. Micciche, Portsmouth, NH: Boynton/Cook Heinemann, 2003, pp. 33–42.
Morgan, Jeffrey. "Emptiness and the Education of the Emotions." *Educational Philosophy and Theory*, vol. 47, no. 3, 2015, pp. 291–304.
Murray, Donald. "Teach Writing as a Process Not Product." *The Essential Don Murray: Lessons from America's Greatest Writing Teacher*, edited by Thomas Newkirk and Lisa C. Miller, Portsmouth, NH: Boynton/Cook Heinemann, 2009.
—. "The Essential Delay: When Writer's Block Isn't." *When a Writer Can't Write*, edited by Mike Rose, New York: Guilford Press, 1985, pp. 219–226.
Musgrove, Laurence E. "Attitudes toward Writing." *JAEPL*, vol. 4, 1998–1999, pp. 1–9.
Ñāṇamolí, Bhikku and Bhikku Bodhi. *The Middle Length Discourses of the Buddha: A Translation of the Majjhima Nikāya*. Somerville, MA: Wisdom Publications, 2015.
"Nandana Sutra: Delight." Translated by Thanissaro Bhikkhu. *Access to Insight (Legacy Edition)*, www.accesstoinsight.org/tipitaka/sn/sn04/sn04.008.than.html.

National Center for Education Statistics. "Back to School Statistics." https://nces.ed.gov/fastfacts/display.asp?id=372.
NCTE. "Professional Knowledge for the Teaching of Writing," 2016, www.ncte.org/positions/statements/teaching-writing.
Nelson, Eric S. "Language and Emptiness in Chan Buddhism and the Early Heidegger." *Journal of Chinese Philosophy*, vol. 37, no. 3, 2010, pp. 472–492.
Newkirk, Thomas. *The Art of Slow Reading: Six Time-Honored Practices for Engagement*. Portsmouth, NH: Heinemann, 2011.
—. "Montaigne's Revisions." *Rhetoric Review*, vol. 24, no. 3, 2005, pp. 298–315.
—. *The School Essay Manifesto: Reclaiming the Essay for Students and Teachers*. Discover Writing Company, 2005.
Newkirk, Thomas and Lisa C. Miller, editors. *The Essential Don Murray: Lessons from America's Greatest Writing Teacher*. Portsmouth, NH: Boynton/Cook Heinemann, 2009.
Nienkamp, Jean. *Internal Rhetorics: Toward a History and Theory of Self-Persuasion*. Carbondale, IL: Southern Illinois University Press, 2001.
Olson, Gary A. "Toward a Post-Process Composition: Abandoning the Rhetoric of Assertion." *Post Process Theory: Beyond the Writing-Process Paradigm*, edited by Thomas Kent, Carbondale, IL: Southern Illinois University Press, 1999.
Ong, Walter J. *Orality and Literacy: The Technologizing of the Word*. Routledge, 1982.
—. "The Writer's Audience Is Always a Fiction." *PMLA*, vol. 90, no. 1, 1975, pp. 9–21.
O'Reilley, Mary Rose. *Radical Presence: Teaching as Contemplative Practice*. Heinemann, 1998.
—. *The Peaceable Kingdom*. Portsmouth, NH: Boynton/Cook Heinemann, 1993.
Papoulis, Irene. "Spirituality and Composition: One Teacher's Thoughts." *The Journal of the Assembly for Expanded Perspectives on Learning*, vol. 2, Winter 1996, pp. 10–17.
Park, Douglas B. "The Meanings of 'Audience.'" *Landmark Essays on Rhetorical Invention in Writing*, edited by Richard E. Young and Yameng Liu, Routledge, 1994, pp. 181–191.
Peary, Alexandria. "The Role of Mindfulness in *Kairos*." *Rhetoric Review*, vol. 35, no. 1, 2016, pp. 22–34.
—. "The Terrain of Prewriting." *Journal of Creative Writing Studies*, vol. 2, no. 1, 2016, http://scholarworks.rit.edu/jcws/vol2/iss1/1.
Perelman, Chaïm and L. Olbrechts-Tyteca. *The New Rhetoric: A Treatise on Argumentation*. University of Notre Dame Press, 1969.
Perl, Sondra. *Felt Sense: Writing with the Body*. Heinemann, 2004.
Perl, Sondra, and Arthur Egendorf. "The Process of Creative Discovery: Theory, Research, and Implications for Teaching." *Language: Linguistics, Stylistics, and the Teaching of Composition*, edited by Donald A. McQuade, Carbondale, IL: Southern Illinois University Press, 1986, pp. 251–268.
Phelps, Louise. *Composition as a Human Science*. New York: Oxford University Press, 1988.
Polanyi, Michael. *The Tacit Dimension*. Chicago: University of Chicago Press, 2009.
Priest, Graham. "The Structure of Emptiness." *Philosophy East & West*, vol. 59, no. 4, 2009, pp. 467–480.
Pullen, Terri G. "Active Receptivity: The Positive, *Mindful* Flow of Mental Energy." *JAEPL*, vol. 3, 1997–1998, pp. 23–31.
Quineau, Raymond. *Exercises in Style*. New York: New Directions, 1981.
Race, William H. "The Word *Kairos* in Greek Drama." *Transactions of the American Philological Association (1974-)*, vol. 111, 1981, pp. 197–213.
Railey, Jennifer McMahon. "Dependent Origination and the Dual-Nature of the Japanese Aesthetic." *Asian Philosophy*, vol. 7, no. 2, 1997, pp. 123–132.

"Rajja Sutra: Rulership." Translated by Thanissaro Bhikkhu. *Access to Insight (Legacy Edition)*, www.accesstoinsight.org/tipitaka/sn/sn04/sn04.020.than.html.

Ratcliffe, Krista. *Rhetorical Listening: Identification, Gender, Whiteness*. Carbondale, IL: Southern Illinois University Press, 2005.

Rechtschaffen, Daniel. *The Mindful Way of Education: Cultivating Well-Being in Teachers and Students*. New York: Norton, 2014.

Reynolds, Nedra. *Geographies of Writing: Inhabiting Places and Encountering Difference*. Carbondale, IL: Southern Illinois University Press, 2007.

Rickert, Thomas J. *Ambient Rhetoric: The Attunements of Rhetorical Being*. Pittsburgh, PA: University of Pittsburgh Press, 2013.

Rogers, Carl R. "Communication: Its Blocking and Its Facilitation." *Etc.*, vol. 9, 1952, pp. 83–88.

Rohman, D. Gordon. "Pre-Writing: The Stage of Discovery in the Writing Process." *College Composition and Communication*, vol. 16, 1965, pp. 106–112.

Rohman, D. Gordon, and Albert O. Wlecke. *Writer's Block: The Cognitive Dimension*. Carbondale, IL: Southern Illinois University Press, 1984.

—. *Pre-Writing: The Construction and Application of Models for Concept Formation in Writing*. Michigan State University, Cooperative Research Project No. 2174, 1964.

Rorty, Richard. *Contingency, Irony, and Solidarity*. Cambridge: Cambridge University Press, 1989.

Rose, Mike. "Rigid Rules, Inflexible Plans, and the Stifling of Language: A Cognitivist Analysis of Writer's Block." *Teaching Composition: Background Readings*, 2nd ed., edited by T. R. Johnson, Boston: Bedford/St. Martin's, 2005, pp. 123–135.

Rose, Mike, editor. *When a Writer Can't Write*. Guilford Press, 1985.

Rotne, Nikolaj Flor, and Didde Flor Rotne. *Everybody Present: Mindfulness in Education*. Berkeley, CA: Parallax Press, 2013.

Ryan, Tim. *A Mindful Nation: How a Simple Practice Can Help Reduce Stress, Improve Performance, and Recapture the American Spirit*. New York: Hay House, 2012.

"Sakalika Sutra: The Stone Sliver." Translated by Thanissaro Bhikkhu. *Access to Insight (Legacy Edition)*, www.accesstoinsight.org/tipitaka/sn/sn04/sn04.013.than.html.

Salzberg, Sharon. *Loving-Kindness: The Revolutionary Art of Happiness*. Boulder, CO: Shambhala Classics, 1995.

Sánchez, Raúl. "Outside the Text: Retheorizing Empiricism and Identity." *College English*, vol. 74, no. 3, 2012, pp. 234–246.

—. *The Function of Theory in Composition Studies*. New York: State University of New York Press, 2005.

Schonert-Reichl, Kimberly A., and Robert W. Roeser. "Mindfulness in Education: Introduction and Overview of the Handbook." *Mindfulness in Behavioral Health*, edited by Kimberly A. Schonert Reichl and Robert W. Roeser, New York: Springer, 2016, pp. 3–16.

Seitz, James. "Roland Barthes, Reading, and Roleplay: Composition's Misguided Rejection of Fragmentary Texts." *College English*, vol. 53, no. 7, November 1991, pp. 815–825.

Shih, Jienshen F. "Buddhist Learning: A Process To Be Enlightened." *Non-Western Perspectives on Learning and Knowing*, edited by Sharan B. Merriam, Malabar, FL: Krieger, 2007, pp. 101–112.

Shonin, Edo, et al. "Mindfulness of Emptiness and the Emptiness of Mindfulness." *Buddhist Foundations of Mindfulness*, edited by Edo Shonin, William Van Gordon, and Nirbhay N. Singh, New York: Springer, 2015, pp. 159–178.

Short, Bryan C. "The Temporality of Rhetoric." *Rhetoric Review*, vol. 7, no. 2, 1989, pp. 367–379.

Shusterman, Richard. *Body Consciousness: A Philosophy of Mindfulness and Somaesthetics*. New York: Cambridge University Press, 2008.

—. "Thinking through the Body, Educating for the Humanities: A Plea for Somaesthetics." *Journal of Aesthetic Education*, vol. 40, no. 1, 2006, pp. 1–21.

Siderits, Mark, and Shōryū Katsura. *Nāgārjuna's Middle Way*. Somerville, MA: Wisdom Publications, 2013.

Siegel, Daniel J. *The Mindful Brain: Reflection and Attunement in the Cultivation of Well-Being*. New York: Norton, 2007.

Siegel, Daniel J., Madeleine W. Siegel, and Suzanne C. Parker. "Internal Education and the Roots of Resilience: Relationships and Reflection as the New R's of Education." *Mindfulness in Behavioral Health*, edited by Kimberly A. Schonert-Reichl and Robert W. Roeser, New York: Springer, 2016, pp. 47–63.

Simmer-Brown, Judith. "Preface to the Vintage Spiritual Classics Edition." *Buddhist Wisdom: The Heart Sutra and The Heart Sutra*. London: Vintage, 2001, pp. xv–xxx.

Sipiora, Phillip, and James S. Baumlin, editors. *Rhetoric and Kairos: Essays in History, Theory, and Praxis*. New York: State University of New York Press, 2002.

Smith, Erec. "Buddhism's Pedagogical Contribution to Mindfulness." *JAEPL*, vol. 21, 2015–2016, pp. 36–46.

Smith, Frank. *Writing and the Writer*. 2nd ed., New Jersey: Lawrence Erlbaum Associates, 1994.

Smith, John E. "Time and Qualitative Time." *The Review of Metaphysics*, vol. 40, no. 1, 1986, pp. 3–16.

Stafford, William. *Writing the Australian Crawl: Views on the Writer's Vocation*. Michigan: University of Michigan Press, 1978.

Stanley, Steven. "From Discourse to Awareness: Rhetoric, Mindfulness, and a Psychology without Foundations." *Theory Psychology*, vol. 23, no. 1, 2012, pp. 60–80.

Steiner, Rudolf. *The Karma of Materialism*. Steiner Books, 1986.

Stewart, Graeme, Tricia Anne Seifert, and Carol Rolheiser. "Anxiety and Self-Efficacy's Relationship with Undergraduate Students' Perceptions of the Use of Metacognitive Writing Strategies." *The Canadian Journal for the Scholarship of Teaching and Learning*, vol. 6, no. 1, 2014, pp. 1–17.

Sun, Jessie. "Mindfulness in Context: A Historical Discourse Analysis." *Contemporary Buddhism*, Vol. 15, No. 2, 2014, pp. 394–415.

Suzuki, Shunryu. *Zen Mind, Beginner's Mind: Informal Talks on Zen Meditation and Practice*. Boston, MA: Shambhala Publications, 1996.

Thera, Nyanaponika. *The Heart of Buddhist Meditation*. New York: Citadel Press, 1962.

Thompson, Evan. *Waking, Dreaming, Being: Self and Consciousness in Neuroscience, Meditation, and Philosophy*. New York: Columbia University Press, 2015.

Tobin, Lad. "Process Pedagogy." *A Guide to Composition Pedagogies*, edited by Gary Tate, Amy Rupiper, and Kurt Schick, New York: Oxford University Press, 2001.

Tompkins, Jane. "Me and My Shadow." *New Literary History*, vol. 19, no. 1, 1987, pp. 169–178.

Toulmin, Stephen. "The Inwardness of Mental Life." *Critical Inquiry*, vol. 6, no. 1, 1979, pp. 1–16.

Tremmel, Robert. *Zen and the Practice of Teaching English*. Portsmouth, NH: Boynton/Cook Heinemann, 1999.

Trimbur, John. "Beyond Cognition: The Voices in Inner Speech." *Rhetoric Review*, vol. 5, no. 2, 1987, pp. 211–221.

Trungpa, Chögyam. *Meditation in Action*. Boulder, CO: Shambhala Classics, 2010.

—. *Cutting through Spiritual Materialism*. Shambhala Classics, 2002.

—. *The Essential Chögyam Trungpa*. Edited by Carolyn Rose Gimian. Boulder, CO: Shambhala Classics, 1999.

Tsering, Geshi Tashi. *Emptiness: The Foundation of Buddhist Thought*. Somerville, MA: Wisdom Publications, 2009.

Tworkov, Helen. "Interbeing with Thich Nhat Hanh: An Interview." *Tricycle*, Summer 1995, https://tricycle.org/magazine/interbeing-thich-nhat-hanh-interview/.

Vatz, Richard E. "The Myth of the Rhetorical Situation." *Philosophy & Rhetoric*, vol. 6, no. 3, 1973, pp. 154–161.

Walshe, Maurice. *The Long Discourses of the Buddha: A Translation of the Dīgha Nikāya*. Somerville, MA: Wisdom Publications, 1995.

Wardle, Elizabeth. " 'Mutt Genres' and the Goal of FYC: Can We Help Students Write the Genres of the University?" *College Composition and Communication*, vol. 60, no. 4, 2009, pp. 765–789.

Watts, Alan W. *The Spirit of Zen: A Way of Life, Work, and Art in the Far East*. Grove Press, 1958.

— . *The Wisdom of Insecurity*. London: Vintage, 1951.

Wenger, Christy I. *Yoga Minds, Writing Bodies: Contemplative Writing Pedagogy*. Anderson, SC: Parlor Press, 2015.

—. "Writing Yogis: Breathing Our Way to Mindfulness and Balance in Embodied Writing Pedagogy." *JAEPL*, vol. 18, 2012–2013, pp. 24–39.

Williams, Bronwyn T. *Literacy Practices and Perceptions of Agency: Composing Identities*. New York: Routledge, 2018.

Wilson, Joel. "Somaesthetics, Composition, and the Ritual of Writing." *Pedagogy*, vol. 15, no. 1, 2015, pp. 173–182.

Winans, Amy E. "Cultivating Critical Emotional Literacy: Cognitive and Contemplative Approaches to Engaging Difference." *College English*, vol. 75, no. 2, 2012, pp. 150–170.

Yagelski, Robert P. *Writing as a Way of Being: Writing Instruction, Nonduality, and the Crisis of Sustainability*. (Research and Teaching in Rhetoric and Composition.) New York: Hampton Press, 2011.

Yancey, Kathleen. *Reflection in the Writing Classroom*. Utah: Utah State University Press, 1998.

Young, Richard E., and Yameng Liu. *Landmark Essays on Rhetorical Invention in Writing*. New York: Routledge, 1994.

Zajonc, Arthur. *Meditation as Contemplative Inquiry: When Knowing Becomes Love*. Massachusetts: Lindisfarne Books, 2009.

INDEX